PVC Pipe—Design and Installation

AWWA MANUAL M23

Second Edition

American Water Works Association

Science and Technology

AWWA unites the drinking water community by developing and distributing authoritative scientific and technological knowledge. Through its members, AWWA develops industry standards for products and processes that advance public health and safety. AWWA also provides quality improvement programs for water and wastewater utilities.

MANUAL OF WATER SUPPLY PRACTICES—M23, Second Edition
PVC Pipe—Design and Installation

Copyright © 2002 American Water Works Association

All rights reserved. No part of this publication may be reproduced or transmitted in any form or by any means, electronic or mechanical, including photocopy, recording, or any information or retrieval system, except in the form of brief excerpts or quotations for review purposes, without the written permission of the publisher.

Library of Congress Cataloging-in-Publication Data has been applied for.

Printed in the United States of America

American Water Works Association
6666 West Quincy Avenue
Denver, CO 80235

ISBN 1-58321-171-3

Printed on recycled paper

Contents

List of Figures, v

List of Tables, vii

Foreword, ix

Acknowledgments, xi

Chapter 1 General Properties of Polyvinyl Chloride Pipe 1

    Background, 1
    Material Properties of PVC Pipe Compounds, 1
    Corrosion, Permeation, and Chemical Resistance, 2
    Environmental Effects, 5

Chapter 2 Testing and Inspection 9

    Testing and Inspection, 9

Chapter 3 Hydraulics 13

    Flow Formulas, 13

Chapter 4 Design Factors Related to External Forces and Conditions 21

    Superimposed Loads, 21
    Flexible Pipe Theory, 24
    Longitudinal Bending, 33
    Expansion and Contraction, 40
    Thrust Restraint—General, 42

Chapter 5 Pressure Capacity 53

    Internal Hydrostatic Pressure, 53
    Distribution Mains, 58
    Transmission Mains, 59
    Injection-Molded PVC Fittings, 62
    Fabricated PVC Fittings, 63
    Dynamic Surge Pressure, 63
    Transmission Pipe Design Example, 67

Chapter 6 Receiving, Storage, and Handling 73

    Receiving, 73
    Storage, 75

Chapter 7 Installation 77

    Scope, 77
    Alignment and Grade, 77
    Installation in Trenches, 77
    Pipe Joints, 82
    Pipe Cutting and Bending, 83
    Pipe Embedment, 84
    Casings, 86

Appurtenances, 88
Thrust Restraint, 91

Chapter 8 Testing and Maintenance **93**
Initial Testing, 93
Timing of the Testing, 93
Initial Cleaning of the Pipeline, 94
Test Preparation, 94
Hydrostatic Testing and Leakage Testing, 94
Test Pressure, 95
Duration of Tests, 95
Allowable Leakage, 96
Disinfecting Water Mains, 96
System Maintenance, 96

Chapter 9 Service Connections **99**
Direct Tapping, 99
Saddle Tapping, 105
Tapping Sleeve and Valve, 107

Appendix A, Chemical Resistance Tables, 109

Appendix B, Flow Friction Loss Tables, 129

Bibliography, 157

Index, 163

List of AWWA Manuals, 167

Figures

1-1 Class 12454 requirements, 4

1-2 Approximate relationship for 12454 PVC for PVC pipe strength properties versus temperature, 6

3-1 Moody diagram—friction factor, 15

3-2 Moody diagram—relative roughness, 16

3-3 Friction loss characteristics of water flow through PVC pipe, 18

3-4 Resistance of valves and fitting to flow of fluids, 19

4-1 Distribution of HS-20 live load through fill, 23

4-2 Bedding angle, 28

4-3 Plasticity chart, 32

4-4 PVC pipe longitudinal bending, 36

4-5 PVC pipe joint offset, 37

4-6 Length variation of unrestrained PVC pipe as a result of temperature change, 41

4-7 Free-body diagram of forces on a pipe bend, 44

4-8 Resultant frictional and passive pressure forces on a pipe bend, 48

4-9 Suggested trench conditions for restrained joints on PVC pipelines, 50

5-1 Stress regression curve for PVC pressure pipe, 55

5-2 Stress regression line, 56

5-3 Strength and life lines of PVC 12454, 57

5-4 Pipeline profile, 67

6-1 Chock block, 75

7-1 Trench cross section showing terminology, 78

7-2 Examples of subditches, 79

7-3 Recommendations for installation and use of soils and aggregates for foundation, embedment, and backfill, 85

7-4 PVC pipe casing skids, 86

7-5 Casing spacer, 87

7-6 Fire hydrant foundation, 90

7-7	Types of thrust blocking, 91
7-8	Types of joint restraint, 92
9-1	Direct PVC pipe tap, 100
9-2	Tapping machine nomenclature, 100
9-3	Cutting/tapping tool, 101
9-4	Mounting the tapping machine, 103
9-5	Cutter feed, 103
9-6	Condition of coupon, 104
9-7	Tapping saddle, 106
9-8	PVC tapping saddle, 107

Tables

1-1 Cell class requirements for rigid poly (vinyl chloride) compounds, 3

4-1 HS-20 and Cooper's E-80 live loads, 24

4-2 PVC pipe stiffness, 26

4-3 Bedding constant values, 28

4-4 Values for the soil support combining factor, S_c, 29

4-5 Values for the modulus of soil reaction $E'b$ for the pipe-zone embedment, psi (MPa), 30

4-6 Soil classification chart (ASTM D2487), 31

4-7 Values for the modulus of soil reaction, E'_n, for the native soil at pipe-zone elevation, 32

4-8 Longitudinal bending stress and strain in PVC pipe at 73.4°F (23°C), 38

4-9 Coefficients of thermal expansion, 40

4-10 Length variation per 10°F (5.6°C) ΔT for PVC pipe, 40

4-11 Estimated bearing strength (undisturbed soil), 45

4-12 Properties of soils used for bedding to calculate F_s and R_s, 50

4-13 In situ values of soil properties for R_s, 52

5-1 Thermal de-rating factors for PVC pressure pipes and fittings, 54

5-2 Pressure classes of PVC pipe (C900), 59

5-3 Pressure ratings of PVC pipe (C905), 60

5-4 Short-term strengths of PVC pipe, 61

5-5 Short-term ratings of PVC pipe, 61

5-6 PVC pressure surge versus DR for 1 ft/sec (0.3 m/sec) instantaneous flow velocity change, 66

7-1 Recommended casing size, 87

7-2 Maximum recommended grouting pressures, 88

8-1 Allowable leakage per 50 joints of PVC pipe, gph, 97

9-1 PVC pipe outside diameters, 102

A-1 Chemical resistance of PVC pressure water pipe, 109

A-2 General chemical resistance of various gasket materials, 114

B-1 Flow friction loss, AWWA C900 PVC pipe, 130

B-2 Flow friction loss, AWWA C905 pipe, 135

B-3 Flow friction loss, ASTM D2241/AWWA C905 pipe, 144

B-4 Flow friction loss, AWWA C909 PVCO pipe, 151

Foreword

This is the second edition of AWWA M23, *PVC Pipe—Design and Installation*. This manual provides the user with both general and technical information to aid in design, procurement, installation, and maintenance of PVC pipe and fittings.

This manual presents a discussion of recommended practices. It is not intended to be a technical commentary on AWWA standards that apply to PVC pipe, fittings, and related appurtenances.

This page intentionally blank.

Acknowledgments

This manual was developed by the AWWA Standards Committee on PVC Pressure Pipe and Fittings. The membership of the committee at the time it approved this manual was as follows:

S.A. McKelvie (Chair), Parsons Brinckerhoff Quade & Douglas, Boston, Mass.

J. Calkins, Certainteed Corporation, Valley Forge, Pa.

J.P. Castronovo, CH2M Hill, Gainesville, Fla.

G.F. Denison, Romac Industries, Inc., Bothell, Wash.

J.L. Diebel, Denver Water, Denver, Colo.

D.L. Eckstein (M23 Subcommittee Chair), The Eckstein Group, Anderson, S.C.

G. Gundel, Specified Fittings, Inc., Bellingham, Wash.

T.H. Greaves, City of Calgary Waterworks, Calgary, Alta.

D.W. Harrington, Bates & Harrington, Inc., Madison Heights, Va.

R. Holme, Earth Tech Canada, Markham, Ont.

J.F. Houle, PW Pipe, Eugene, Ore.

L.A. Kinney, Jr., Bureau of Reclamation, Denver, Colo.

J.H. Lee, Dayton & Knight Ltd., W. Vancouver, B.C.

G.J. Lefort, IPEX Inc., Langley, B.C.

M.D. Meadows (Standards Council Liaison), Brazos River Authority, Waco, Texas

E.W. Misichko, Underwriters Laboratories Inc., Northbrook, Ill.

J.R. Paschal, NSF International, Ann Arbor, Mich.

S. Poole, Epcor Water Services, Edmonton, Alta.

J.G. Richard, Jr., Baton Rouge, La.

J. Riordan, HARCO Fittings, Lynchburg, Va.

E.E. Schmidt, Diamond Plastics Corporation, Grand Island, Neb.

T. Shellenbarger, Dresser Mgf. Div., Dresser Ind., Bradford, Pa.

J.K. Snyder, Snyder Environ. Engrg. Assocs., Audubon, Pa.

J.S. Wailes (Staff Advisor), AWWA, Denver, Colo.

R.P. Walker, Uni-Bell PVC Pipe Association, Dallas, Texas

W.R. Whidden, Post Buckly Schuh & Jernigan, Orlando, Fla.

D.R. Young, Florida Cities Water Co., Sarasota, Fla.

K. Zastrow, Underwriters Laboratories Inc., Northbrook, Ill.

Credit is extended to the Uni-Bell PVC Pipe Association, Dallas, Texas, for granting permission to reprint many of the graphics and tables from the Uni-Bell *Handbook of PVC Pipe: Design and Construction*, copyright 2001.

This page intentionally blank.

AWWA MANUAL M23

Chapter 1

General Properties of Polyvinyl Chloride Pipe

BACKGROUND

Polyvinyl chloride (PVC) was discovered in the late nineteenth century. Scientists at that time found the new plastic material unusual in that it appeared nearly inert to most chemicals. However, it was soon discovered that the material was resistant to change, and it was concluded that the material could not be easily formed or processed into usable applications.

In the 1920s, scientific curiosity again brought polyvinyl chloride to public attention. In Europe and America, extended efforts eventually brought PVC plastics to the modern world. Technology, worldwide and particularly in Germany, slowly evolved for the use of PVC in its unplasticized, rigid form, which today is used in the production of a great many extruded and molded products. In the mid-1930s, German scientists and engineers developed and produced limited quantities of PVC pipe. Some PVC pipe installed at that time continues to provide satisfactory service today. Molecularly oriented polyvinyl chloride (PVCO) pressure pipe has been installed in Europe since the early 1970s and in North America since 1991.

MATERIAL PROPERTIES OF PVC PIPE COMPOUNDS

Polyvinyl chloride pipe and fabricated fittings derive properties and characteristics from the properties of their raw material components. Essentially, PVC pipe and fabricated fittings are manufactured from PVC extrusion compounds. Injection molded fittings use slightly different molding compounds. PVCO is manufactured from conventional PVC extrusion compounds. The following summary of the material properties for these compounds provides a solid foundation for an understanding and appreciation of PVC pipe properties.

Polyvinyl chloride resin, the basic building block of PVC pipe, is a polymer derived from natural gas or petroleum, salt water, and air. PVC resin, produced by any of the common manufacturing processes (bulk, suspension, or emulsion), is combined with heat stabilizers, lubricants, and other ingredients to make PVC compound that can be extruded into pipe or molded into fittings.

Chemical and taste-and-odor evaluations of PVC compounds for potable water conveyance are conducted in accordance with procedures established by NSF International.* The extracted water must not exceed the maximum contaminant levels established by the US Environmental Protection Agency's (USEPA) National Interim Primary Drinking Water Regulations (1975) and by the NSF limits of acceptance for residual vinyl chloride monomer and for taste and odor as shown in Table 1-1 of NSF Standard 61. Monitoring is conducted by NSF International or approved laboratories.

PVC pipe extrusion compounds must provide acceptable design stress properties as determined by long-term testing under hydrostatic pressure. Hydrostatic design stress ratings for pipe compounds are established after 10,000 hr of hydrostatic testing. Long-term performance of injection molded PVC fittings compounds are subject to at least 2,000 hr of hydrostatic testing.

AWWA's PVC pipe and fittings standards define the basic properties of PVC compound, using the American Society for Testing and Materials (ASTM) Specification D1784, *Standard Specification for Rigid Poly (Vinyl Chloride) (PVC) Compounds and Chlorinated Poly (Vinyl Chloride) (CPVC) Compounds*. The specification includes a five-digit cell class designation system by which PVC compounds are classified according to their physical properties.

As shown in Table 1-1, the five properties designated are (1) base resin, (2) impact strength, (3) tensile strength, (4) elastic modulus in tension, and (5) deflection temperature under loading. Figure 1-1 shows how the classification system establishes minimum properties for the compound 12454, which is used in PVC pressure pipe manufactured in accordance with AWWA C900, *Polyvinyl Chloride (PVC) Pressure Pipe and Fabricated Fittings, 4 In. Through 12 In. (100 mm Through 300 mm), for Water Distribution;*† AWWA C905, *Polyvinyl Chloride (PVC) Pressure Pipe and Fabricated Fittings, 14 In. Through 48 In. (350 mm Through 1,200 mm), for Water Transmission and Distribution;*† and AWWA C909, *Molecularly Oriented Polyvinyl Chloride (PVCO) Pressure Pipe, 4 In. Through 12 In. (100 mm Through 300 mm), for Water Distribution*. The material classification can be found on the pipe as part of its identification marking.

Many of the important properties of PVC pipe are predetermined by the characteristics of the PVC compound from which the pipe is extruded. PVC pressure pipe manufactured in accordance with AWWA C900, C905, or C909 must be extruded from PVC compound with cell classification 12454-B or better. Those compounds must also qualify for a hydrostatic design basis of 4,000 psi (27.58 MPa) for water at 73.4°F (23°C) per the requirements of PPI† TR-3.

The manner in which selected materials are identified by this classification system is illustrated by a Class 12454 rigid PVC compound having the requirements shown in Table 1-1 and Figure 1-1.

CORROSION, PERMEATION, AND CHEMICAL RESISTANCE

PVC and PVCO pipes are resistant to almost all types of corrosion both chemical and electrochemical that are experienced in underground piping systems. Because PVC is a nonconductor, galvanic and electrochemical effects are nonexistent in PVC piping systems. PVC pipe cannot be damaged by aggressive waters or corrosive soils. Consequently, no lining, coating, cathodic protection, or plastic encasement is required when PVC and PVCO pipes are used.

---

*NSF International, 789 N. Dixboro Rd., Ann Arbor, MI 48105.
†Plastics Pipe Institute, 1275 K St. N.W., Suite 400, Washington, D.C. 20005.

Table 1-1 Cell class requirements for rigid poly (vinyl chloride) compounds*

Order No.	Designation Property and Unit	Cell Limits							
		0	1	2	3	4	5	6	7	8
1	Base resin	Unspecified	Poly (vinyl chloride) homopolymer	Chlorinated poly (vinyl chloride)	Ethylene vinyl chloride copolymer	Propylene vinyl chloride copolymer	Vinyl acetate-vinyl chloride copolymer	Alkyl vinyl ether-vinyl chloride copolymer		
2	Impact strength (Izod) min.									
	J/m of notch	Unspecified	<34.7	34.7	80.1	266.9	533.8	800.7		
	ft-lb/in. of notch		<0.65	0.65	1.5	5.0	10.0	15.0		
3	Tensile strength, min:									
	MPa	Unspecified	<34.5	34.5	41.4	48.3	55.2			
	psi		<5,000	5,000	6,000	7,000	8,000			
4	Modulus of elasticity in tension, min:									
	MPa	Unspecified	<1,930	1,930	2,206	2,482	2,758	3,034		
	psi		<280,000	280,000	320,000	360,000	400,000	440,000		
5	Deflection temperature under load, min. 1.82 MPa (264 psi):									
	deg C	Unspecified	<55	55	60	70	80	90	100	110
	deg F		<131	131	140	158	176	194	212	230

Source: ASTM D1784, American Society for Testing and Materials, 100 Barr Harbor Dr., West Conshohocken, PA 19428-2959.

* The minimum property value will determine the cell number although the maximum expected value may fall within a higher cell.

Note: Flammability. All compounds covered by this specification, when tested in accordance with method D635, shall yield the following results: average extent of burning of <25 mm; average time of burning of <10 sec.

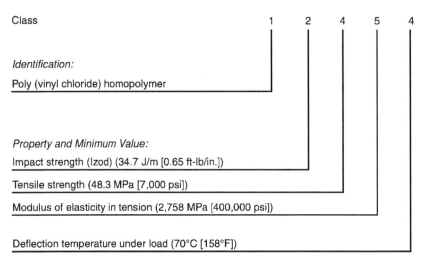

Source: ASTM D1784, American Society for Testing and Materials, 100 Barr Harbor Dr., West Conshohocken, PA 19428-2959.

Note: The cell-type format provides the means for identification and close characterization and specification of material properties, alone or in combination, for a broad range of materials. This type format, however, is subject to possible misapplication since unobtainable property combinations can be selected if the user is not familiar with commercially available materials. The manufacturer should be consulted.

Figure 1-1   Class 12454 requirements

Permeation

The selection of materials is critical for water service and distribution piping in locations where the pipe may be exposed to significant concentrations of pollutants comprised of low molecular weight petroleum products or organic solvents or their vapors. Research has documented that pipe materials, such as polyethylene, polybutylene, polyvinyl chloride, and asbestos cement, and elastomers, such as those used in jointing gaskets and packing glands, may be subject to permeation by lower molecular weight organic solvents or petroleum products. If a water pipe must pass through an area subject to contamination, the manufacturer should be consulted regarding permeation of pipe walls, jointing materials, etc., *before* selecting materials for use in that area.

Chemical Resistance

Pipe. Response of PVC pipe under normal conditions to commonly anticipated chemical exposures is shown in Table A-1 in Appendix A. Resistance of PVC pipe to reaction with or attack by the chemical substances listed has been determined by research and investigation. The information is primarily based on the immersion of unstressed strips into the chemicals and, to a lesser degree, on field experience. In most cases, the detailed test conditions, such as stress, exposure time, change in weight, change in volume, and change in strength, were not reported. Because of the complexity of some organochemical reactions, additional long-term testing should be performed for critical applications. Data provided are intended only as a guide and should not necessarily be regarded as applicable to all exposure durations, concentrations, or working conditions. This chemical resistance data is similar for PVCO pipe.

Gaskets. A check of the chemical resistance of the gasket should be completed independently of that for the pipe. Because gasket and pipe materials are different, so too are their abilities to resist chemical attack. Similarly, charts for resistance of gasket materials to chemical attack are based on manufacturers' testing and experience. The use of these charts is complicated by the fact that more than one elastomer may be present in a rubber compound. Chemical resistance information for commonly used gasket materials is provided in Table A-2 in Appendix A.

Table A-2 is a general guide to the suitability of various elastomers currently used in these chemicals and services. The ratings are primarily based on literature published by various polymer suppliers and rubber manufacturers, as well as the opinions of experienced compounders. Several factors must be considered in using a rubber or polymer part. The most important of these factors include the following:

- Temperature of service. Higher temperatures increase the effect of all chemicals on polymers. The increase varies with the polymer and the chemical. A compound quite suitable at room temperature may perform poorly at elevated temperatures.
- Conditions of service. A compound that swells considerably might still function well as a static seal yet fail in any dynamic application.
- Grade of the polymer. Many types of polymers are available in different grades that vary greatly in chemical resistance.
- Compound itself. Compounds designed for other outstanding properties may be poorer in performance in a chemical than one designed especially for fluid resistance.
- Availability. Consult the elastomer manufacturers for availability of a compound for use as a PVC pipe gasket material.

If it is anticipated that gasket elastomers will be exposed to aggressive chemicals, it is advisable to test the elastomers.

ENVIRONMENTAL EFFECTS

The following paragraphs discuss the effects of environmental factors on PVC pipe, including temperature, biological attack, weather, abrasion, and tuberculation.

Thermal Effects

The performance of PVC pipe is significantly related to its operating temperature. Because it is a thermoplastic material, PVC will display variations in its physical properties as temperature changes (Figure 1-2). PVC pipe can be installed properly over the ambient temperature range in which construction crews can work. PVC pipe is rated for performance properties at a temperature of 73.4°F (23°C); however, it is recognized that operating temperatures of 33–90°F (1–32°C) do exist in water systems. As the operating temperature decreases, the pipe's stiffness and tensile strength increase, thereby increasing the pipe's pressure capacity and its ability to resist earth-loading deflection. At the same time, PVC pipe loses impact strength and becomes less ductile as temperature decreases, necessitating greater handling care in sub-zero weather. As the operating temperature increases, the impact strength and flexibility of PVC pipe increases. However, with the increase in temperature, PVC pipe loses tensile strength and stiffness; consequently, the pressure capacity of the pipe will be reduced and more care will be needed during installation to avoid excessive deflection.

Most municipal water systems operate at temperatures at or below 73.4°F (23°C). In these applications, the actual pressure capacity of PVC pipe will be equal to or greater than the product's rated pressure. Intermittent water system temperatures above 73.4°F (23°C) do not warrant derating of pipe or fitting pressure designations.

New users and installers of PVC pipe should be aware of the pipe's capacity to expand and contract in response to changes in temperature. The PVC coefficient of thermal expansion is roughly five times the normal value for cast iron or steel. Provisions must be made in design and installation to accommodate expansion and contraction if the pipeline is to provide service over a broad range of operating temperatures. In general, allowance must be made for 3/8 in. of expansion or contraction for every

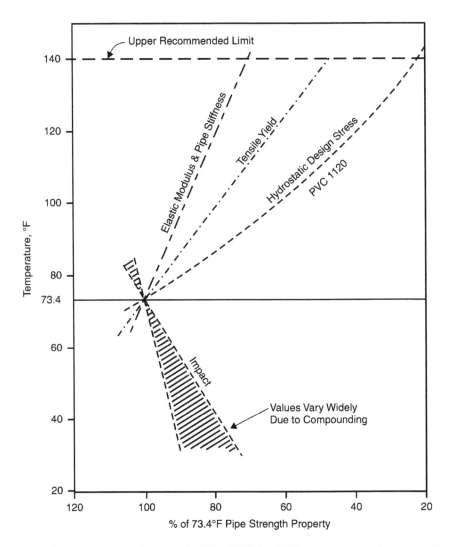

Figure 1-2 Approximate relationship for 12454 PVC for PVC pipe strength properties versus temperature

100 ft (30.5 m) of pipe for each 10°F (5.6°C) change in temperature. Gasketed joints provide excellent allowance for thermal expansion and contraction of PVC pipelines. The coefficient of thermal expansion for PVCO is the same as for PVC.

Resistance to Biological Attack

PVC pipe is nearly totally resistant to biological attack. Biological attack can be described as degradation or deterioration caused by the action of living microorganisms or macroorganisms. Microorganisms that attack organic materials are normally listed as fungi and bacteria. Macroorganisms that can affect organic materials located underground include an extremely broad category of living organisms; for example, grass roots, termites, and rodents. The performance of PVC pipe in environments providing severe exposure to biological attack in its various anticipated forms has been studied and evaluated since the 1930s.

PVC pipe will not deteriorate or break down under attack from bacteria or other microorganisms, nor will it serve as a nutrient to microorganisms, macroorganisms, or fungi. No cases have been documented where buried PVC pipe products have degraded or deteriorated because of biological action. As a result, no special engineering or installation procedures are presently required to protect PVC or PVCO pipe from any known form of biological attack.

Elastomeric seals are manufactured from organochemical materials, which can be formulated to produce a variety of properties. Some elastomers are susceptible to biological attack, whereas others provide resistance comparable to those inherent in polyvinyl chloride. PVC pipe manufacturers select gaskets produced from elastomeric compounds that provide high resistance. A material that will not support bacterial growth is a requirement, particularly in potable water systems.

In normal practice, when installing PVC pipe with gasketed joints, assembly of joints is facilitated using a lubricant applied in accordance with the manufacturer's instructions. Care must be exercised in the selection of this lubricant to ensure compatibility with the elastomeric seal and the PVC pipe and to ensure that the lubricant will not support the growth of fungi or bacteria. Care must also be taken to ensure that only the amount of lubricant required to facilitate assembly is used. Excess lubricant can adversely affect water quality and ultimately delay commissioning of a water system. *Only the lubricant recommended by the pipe manufacturer should be used.* These lubricants must also satisfy all NSF 61 requirements.

Weathering Resistance

PVC pipe can incur surface damage when subjected to long-term exposure to ultraviolet (UV) radiation from sunlight. This effect is called ultraviolet degradation. Unless specifically formulated to provide substantial protection from UV radiation (for example, PVC house siding), or unless a limited service life is acceptable, PVC pipe is not recommended for applications where it will be continuously exposed to direct sunlight without some form of physical protection (such as paint or wrapping).

Ultraviolet degradation in PVC occurs when energy from the UV radiation causes excitation of the molecular bonds in the plastic. The resulting reaction occurs only on the exposed surface of PVC pipe and penetrates the material less than 0.001 in. (0.025 mm). Within the affected zone of reaction, the structure of the PVC molecule is permanently altered with the molecules being converted into a complex structure typified by polyene formations. The polyene molecule causes a light yellow coloration on the PVC pipe and slightly increases its tensile strength.

Regarding the organochemical reactions that characterize ultraviolet deterioration of PVC, the following should be noted:

- UV degradation results in color change, slight increase in tensile strength, slight increase in the modulus of tensile elasticity, and decrease in impact strength in PVC pipe.
- UV degradation does not continue when exposure to UV radiation is terminated.
- UV degradation occurs only in the plastic material directly exposed to UV radiation and to an extremely shallow penetration depth.
- UV degradation of PVC pipe formulated for buried use will not have significant adverse effect with up to two full years of outdoor weathering and direct exposure to sunlight.

The above is also true in regard to PVCO pipe.

Abrasion

After years of investigation and observation, it has been established that the combination of PVC resin, extenders, and various additives in PVC compounds, plus the methods of extrusion for PVC pipe, produce a resilient product with good resistance to abrasive conditions.

Many investigations and tests have been conducted, both in North America and Europe, by manufacturers, independent laboratories, and universities seeking to define PVC pipe's response to abrasion. Although the approaches to the various tests and investigations have varied substantially, the data developed has been consistent in defining the extent of PVC pipe resistance to abrasion. The nature and resiliency of PVC pipe cause it to gradually erode over a broad area when exposed to extreme abrasion, rather than to develop the characteristic localized pitting and more rapid failure observed in pipe products with lower abrasion resistance.

PVC pipe is well suited to applications where abrasive conditions are anticipated. In extremely abrasive exposures, wear must be anticipated; however, in many conditions PVC pipe can significantly reduce maintenance costs incurred because of extreme abrasion. It should be noted that potable water, regardless of its makeup, is not considered abrasive to PVC pipe.

Tuberculation

Soluble encrustants (such as calcium carbonate) in some water supplies do not precipitate onto the smooth walls of PVC or PVCO pipe. Because these materials do not corrode, there is no tuberculation caused by corrosion by-products.

AWWA MANUAL M23

Chapter 2

Testing and Inspection

The technology of PVC pipe manufacturing processes is extensive and involved. It may be traced from oil or gas wells through petrochemical plants to the PVC compounding operations and finally to the automated extrusion, molding, and fabrication operations before a finished PVC product is ready for testing, inspection, and delivery. This chapter covers testing and inspection as it applies to the manufacturing of PVC and PVCO pipe products.

TESTING AND INSPECTION

Testing and inspection in PVC pipe manufacturing may be divided into three categories: (1) qualification testing, (2) quality control testing, and (3) assurance testing.

Qualification Testing

Qualification testing is performed on piping products and on the materials from which they are produced to ensure that the finished products meet the requirements of applicable specifications. Qualification testing must demonstrate that the materials, process equipment, and manufacturing technology consistently yield, through proper production procedures and controls, finished products that comply with applicable standards.

The following qualification tests are required in the manufacture of AWWA C900, C905, and C909 PVC pipe to evaluate the design properties noted.

PVC extrusion compound cell classification testing. This qualification test, as defined in ASTM D1784, is required and performed to establish primary mechanical and chemical properties of the PVC material from which the finished pipe products are produced.

Gasketed joint design testing. One option for testing joint design is to perform pressure tests to verify that joint assemblies qualify for a hydrostatic design basis category of 4,000 psi (27.6 MPa).

Toxicological testing. This qualification test is performed to verify that metals and chemicals cannot be extracted by water in quantities termed toxic, carcinogenic, teratogenic, or mutagenic, which produce adverse physiological effects in humans. The test, as specified in ANSI/NSF 61, is required for all PVC potable-water piping materials and products.

Long-term hydrostatic strength testing. This qualification test is required and performed to establish the maximum allowable design (tensile) stress in the wall of PVC pipe in a circumferential orientation (hoop stress) as a result of internal pressure applied continuously with a high level of certainty that failure of the pipe cannot occur.

Joint performance testing. This qualification test is performed to verify a leak-free design of a specified pipe joint that will maintain a proper connection and seal.

Lap-shear test. This test is used to verify that fabricated-fitting solvent-cementing procedures result in minimum average lap-strengths of 900 psi (6.2 MPa). Lap-sheer test samples are produced by solvent-cementing of component pipe segments identical to those that are used to fabricate fittings.

Quality Control Testing

Quality control testing is routinely performed on specimens of PVC piping products as they are manufactured to ensure that the products comply with applicable standards. Quality control testing includes, but is not limited to, inspection and testing to verify proper dimensional, physical, and mechanical properties. Frequently, quality control tests are required that may not define a desired finished product property but that do verify the use of proper procedures and controls in the manufacturing process. Quality control tests and inspection required in the manufacture of AWWA C900, C905 PVC, and C909 PVCO products are as follows.

Workmanship inspection. Inspection is conducted to ensure that the PVC pipe product is homogeneous throughout free from voids, cracks, inclusions, and other defects and reasonably uniform in color, density, and other physical properties. Surfaces are inspected to ensure that they are free from nicks, gouges, severe scratches, and other such blemishes. Joining surfaces shall be ensured freedom from damage and imperfections.

Marking inspection. Inspection verifies proper marking of the pipe as required in the applicable product standard. Marking of AWWA C900, C905 PVC, and C909 PVCO pipe includes the following:

- PVC or PVCO
- Manufacturer's name or trademark and production-record code
- Nominal pipe size
- Outside diameter regimen (C905 only)
- Dimension ratio (for example, DR 25)
- AWWA pressure class or pressure rating (for example, PC 100)
- AWWA standard designation (for example, AWWA C900)
- Seal of the testing agency that verified the suitability of the pipe material for potable-water service (optional)

Dimension measurement. Measurement of dimensions on a regular and systematic basis is essential. Failure to meet dimensional requirements may render the product unsatisfactory regardless of success in other inspections and tests. All dimensional measurements are made in accordance with ASTM D2122 and include the following:

- Product diameter
- Product wall thickness
- Bell joint dimensions
- Fabricated-fitting configurations
- Length

Some dimensional requirements are defined in the manufacturer's product specifications. Markings of machined couplings and fabricated fitting includes the following:

- Nominal size and deflection angle (if applicable)
- PVC
- AWWA pressure class or pressure rating
- AWWA standard designation
- Manufacturer's name or trademark
- Seal of the testing agency that verified the suitability of the PVC material for potable-water service (optional)

Product packaging inspection. The finished package of PVC pipe prepared for shipment to the customer is inspected to ensure correct pipe quantity and adequate protection of the pipe.

Quick-burst test. The PVC pipe sample is pressurized to burst within the test time period of 60–70 sec. Burst pressure measured must not be less than minimum burst pressure requirements defined in AWWA C900 or C909. Quick-burst testing is conducted in accordance with ASTM D1599. This test is also performed on machined couplings.

Flattening test. The PVC pipe specimen is partially flattened between moving parallel plates. When the pipe is flattened 60 percent (the distance between the parallel plates equals 40 percent of the original outside diameter), the specimen should display no evidence of splitting, cracking, or breaking.

Extrusion quality test. The PVC pipe specimen is immersed in anhydrous (dry) acetone for 20 min. When removed from the acetone bath, the pipe specimen should pass the failure criteria in ASTM D2152. Extrusion quality testing is conducted in accordance with ASTM D2152 and distinguishes only between unfused and properly fused PVC pipe.

Quality control inspection and testing must not be confused with field acceptance testing. Quality control testing is only appropriate during or immediately following the manufacturing process.

Arc test for fabricated fittings. The arc test is required for butt-fused or thermally welded joints in fabricated fittings. Any discontinuity in a segment joint is indicated by the presence of an arc (spark) from a probe tip and is cause for rejection of the fitting.

Fabricated-fitting pressure test. In this test, the fabricated fitting must not fail, balloon, burst, or weep when subjected to an internal pressure test. For C900 fabricated fittings, the internal pressure test is equal to four times its designated pressure class for a minimum of one hour. For C905 fabricated fittings, the internal pressure test is equal to two times its designated pressure rating for a minimum of two hours.

Assurance Testing

Assurance testing is performed at the completion of the manufacturing process to assure the finished products consistently and reliably satisfy the requirements of applicable standards. Quality assurance tests required in the manufacture of AWWA C900, C905 PVC, and C909 PVCO products are as follows.

Sustained pressure test. C900 pipe or fabricated fittings shall not fail, balloon, burst, or weep, as defined in ASTM D1598 at the applicable sustained pressure when tested for 1,000 hr as specified in ASTM D2241.

Hydrostatic proof test. The hydrostatic proof test is required in the manufacture of PVC pipe and machined couplings in accordance with AWWA C900, C905, and C909. In the test, every coupling and piece of PVC or PVCO pipe is proof-tested for a minimum dwell time of 5 sec. C900 and C909 require the hydrostatic proof test be conducted at four times the pressure class (i.e., 4 × 150 psi = 600 psi for DR 18 pipe). C905 requires that the hydrostatic proof test be conducted at two times the pressure rating of the pipe (i.e., 2 × 235 psi = 470 psi for DR 18 pipe).

Hydrostatic Proof Testing

Proof-test frequency may be modified by agreement between manufacturer and producer/supplier.

AWWA MANUAL M23

Chapter 3

Hydraulics

Many empirical formulas and equations have been developed to provide a solution to the problem of flow in pipes and are used daily by water utility engineers. Relatively few specific problems in pipe hydraulics, such as laminar flow, can be solved entirely by rational mathematical means. Solutions to the majority of flow problems depend on experimentally determined coefficients and relationships. Commonly used flow formulas have been developed through research by Darcy, Chezy, Kutter, Scobey, Manning, Weisbach, Hazen, and Williams.

FLOW FORMULAS

Hydraulic flow research and analysis have established that the Hazen-Williams equation can be used for PVC pressure piping system design. Flow conditions may also be analyzed more precisely and with more detail using the Darcy-Weisbach equation.

Darcy-Weisbach Equation

The Darcy-Weisbach equation provides the hydraulic design of PVC pressure water pipe. Relative pipe roughness (ε/D) and Reynolds Number ($R_e = VD/\nu$) are also defined. The commonly used form of the Darcy-Weisbach formula is shown in Eq 3-1.

$$h_f = f \frac{L V_f^2}{D \; 2g} \qquad (3\text{-}1)$$

Where:

h_f = head loss, ft of H_2O
f = friction factor
L = pipe length, ft
D = pipe inside diameter, ft
V_f = mean flow velocity, ft/sec
g = acceleration of gravity, 32.2 ft/sec^2

Investigation and analysis by Neale and Price have established that the friction factor f for hydraulically smooth flow in PVC pipe may be defined by the following equation:

$$\frac{1}{\sqrt{f}} = 2 \log_{10}(R_e \sqrt{f}) - 0.8 \tag{3-2}$$

Where:

f = friction factor
R_e = Reynolds Number

The calculations for the friction factor (f) may be tedious. In common practice, the factor is established by using the Moody diagram shown in Figure 3-1. Relative roughness (ε/D) is related to friction factor (f) as shown in Eq 3-3. Figure 3-2 provides values for relative roughness (ε/D) for various pipe products.

$$\frac{1}{\sqrt{f}} = 1.14 - 2 \log_{10}\left(\frac{\varepsilon}{D} + \frac{9.35}{R_e \sqrt{f}}\right) \tag{3-3}$$

Where:

f = friction factor
ε = 0.000005 ft, PVC pipe
D = pipe inside diameter, ft
R_e = Reynolds Number

Hazen–Williams Equation

The Hazen–Williams flow equation is the most widely accepted and used for calculating pressure pipe flow conditions. The equation can be expressed in the following ways depending on the solution needed. Flow velocity in a pipeline can be calculated using Eq 3-4.

$$V = 1.318 C (R_H)^{0.63} (S)^{0.54} \tag{3-4}$$

Where:

V = flow velocity, ft/sec
C = flow coefficient
R_H = hydraulic radius, ft
   Note: $R_H = \frac{1}{4}(D)$ for pipe flowing full
S = hydraulic slope, ft/ft

Flow rate in gpm, given pressure drop in psi, can be calculated using Eq 3-5.

$$Q = 0.442 d_i^{2.63} C \left(\frac{P_1 - P_2}{L}\right)^{0.54} \tag{3-5}$$

Where:

Q = flow rate, gpm (All gallons are US gallons unless otherwise noted.)
d_i = pipe inside diameter, in.
C = flow coefficient
P_1, P_2 = gauge pressures, psi
L = pipe length, ft

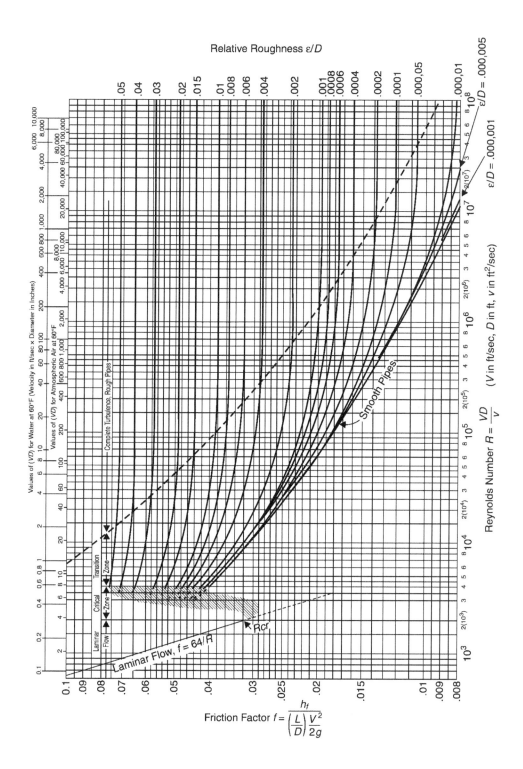

Source: American Society of Mechanical Engineers, New York, NY, *Transactions*, ASME, Vol. 66 (1944) L.F. Moody.

Figure 3-1 Moody diagram—friction factor

16 PVC PIPE—DESIGN AND INSTALLATION

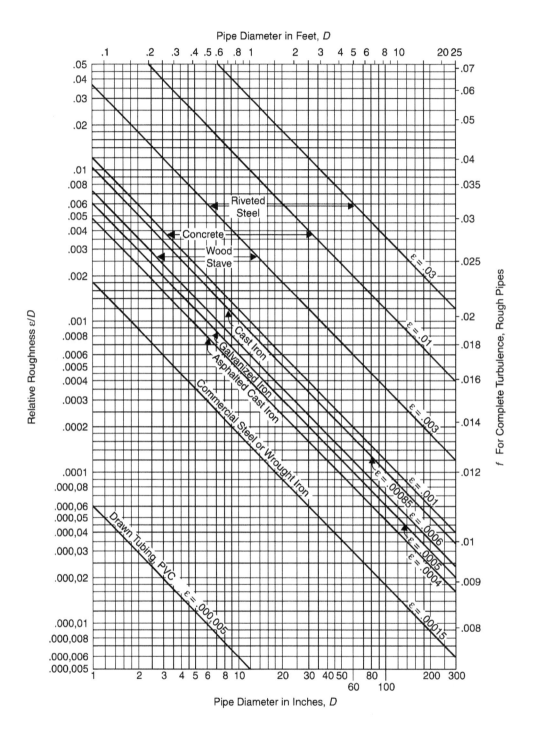

Source: American Society of Mechanical Engineers, New York, NY, *Transactions*, ASME, Vol. 66 (1944) L.F. Moody.

Figure 3-2 Moody diagram—relative roughness

Using Eq 3-6, flow rate can be derived from pressure drop expressed in terms of feet per 1,000 ft.

$$Q = 0.006756 \, C d_i^{2.63} H^{0.54} \qquad (3\text{-}6)$$

Where:

Q = flow rate, gpm
C = flow coefficient
d_i = pipe inside diameter, in.
H = head loss, ft/1,000 ft

Friction loss in hydraulic flow can be derived from Eq 3-7.

$$f = 0.2083 \left(\frac{100}{C}\right)^{1.852} \frac{Q^{1.852}}{d_i^{4.8655}} \qquad (3\text{-}7)$$

Where:

f = friction loss, ft of water/100 ft
C = flow coefficient
Q = flow rate, gpm
d_i = pipe inside diameter, in.

Flow coefficients for PVC pipe have been derived through research and analysis by various researchers, including Neale, Price, Jeppson, and Bishop. Research has established that the Hazen–Williams flow coefficient C can range in value from 155 to 165 for both new and previously used PVC pipe. Therefore, a flow coefficient of $C = 150$ is generally used as a conservative value for the design of PVC piping systems.

Using $C = 150$ for PVC pipe, Eq 3-4 through 3-7 can be simplified as follows for use in designing PVC piping systems:

$$V = 197.7 \, R_H^{0.63} (S)^{0.54} \qquad (3\text{-}8)$$

$$Q = 66.3 \, d_i^{2.63} \left(\frac{P_1 - P_2}{L}\right)^{0.54} \qquad (3\text{-}9)$$

$$Q = 1.0134 \, d_i^{2.63} H^{0.54} \qquad (3\text{-}10)$$

$$f = 0.0984 \, \frac{Q^{1.852}}{d_i^{4.8655}} \qquad (3\text{-}11)$$

Where:

V = flow velocity, ft/sec
R_H = hydraulic radius, ft
S = hydraulic slope, ft/ft
Q = flow rate, gpm
d_i = pipe inside diameter, in.
P_1, P_2 = gauge pressures, psi
L = pipe length, ft
H = head loss, ft/1,000 ft
f = friction loss, ft of water/100 ft

For convenience in design, Tables B-1, B-2, B-3, and B-4 in Appendix B have been developed, based on the Hazen–Williams formula with $C = 150$, to provide flow capacity (gpm), friction loss (feet per 100 ft), and flow velocity (ft/sec) for AWWA C900, AWWA C905, ASTM D2241, and C909 PVC pressure pipe products. Nomographs for solving head loss characteristics are provided in Figures 3-3 and 3-4.

18   PVC PIPE—DESIGN AND INSTALLATION

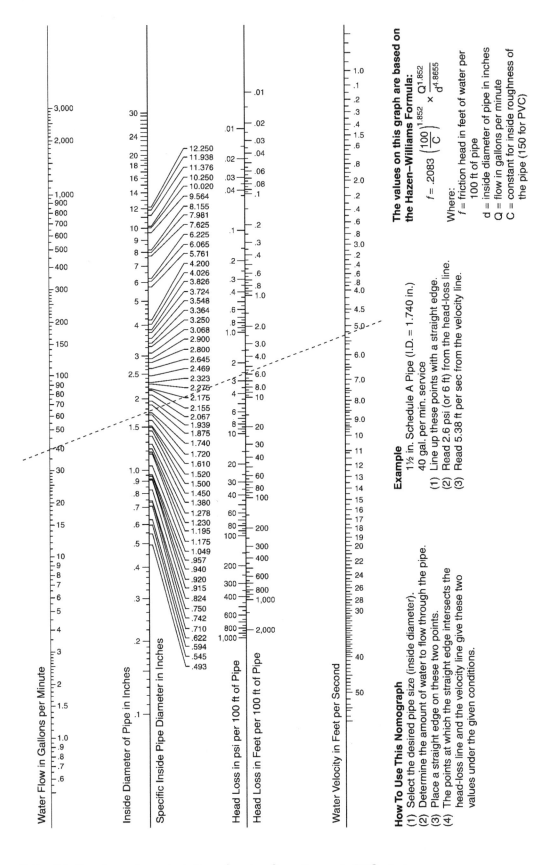

Figure 3-3   Friction loss characteristics of water flow through PVC pipe

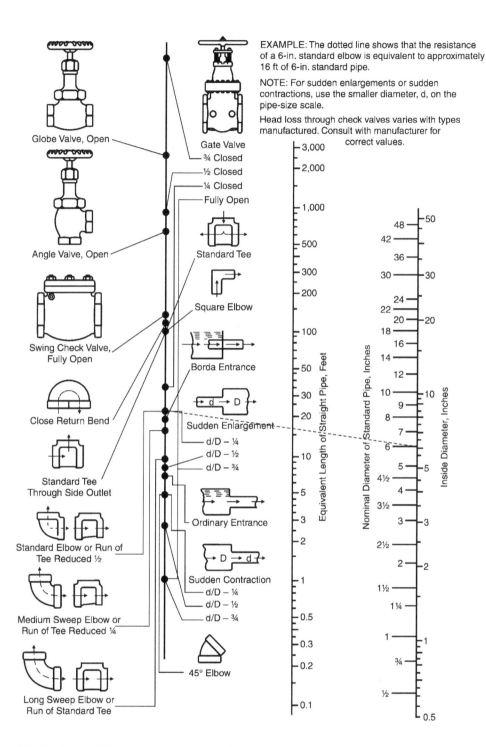

Source: Flow of Fluids through Valves, Fittings, and Pipe. Copyright 1942 by Crane Company.

Figure 3-4 Resistance of valves and fitting to flow of fluids

This page intentionally blank.

AWWA MANUAL M23

Chapter 4

Design Factors Related to External Forces and Conditions

This chapter discusses design considerations related to (1) superimposed loads on buried PVC pipe, (2) flexible pipe theory, (3) longitudinal bending, and (4) thermal expansion and contraction.

SUPERIMPOSED LOADS

Superimposed loads on buried PVC pipe fall into three categories: (1) static earth loads, (2) other dead or static loads, and (3) live loads. In the design of any buried pipe system, all categories of superimposed loads must be considered. In accordance with common design practice, treatment of superimposed loads will consider dead (static) loads and live (dynamic) loads as separate design parameters.

Static Earth Loads

The first solution to the problem of soil-induced loads on buried pipe was published by Professor Anson Marston at Iowa State University in 1913. The Marston loads theory on underground conduits is considered state of the art in determining loading on buried pipe, especially for rigid conduits. Much of the work done on earth-loading technology for buried conduits throughout the world is related, in part, to Marston's load theory, which follows:

Rigid Pipe: $$W_c = C_d\, wB_d \cdot B_d \tag{4-1}$$

Flexible Pipe: $$W_c = C_d\, wB_d \cdot B_c \tag{4-2}$$

Where:

W_c = load on conduit, lb/lin ft
C_d = load coefficient for conduits installed in trenches
w = unit weight of backfill, lb/ft^3
B_d = horizontal width of trench at top of conduit, ft
B_c = horizontal width of conduit, ft

Equations 4-1 and 4-2 are used to calculate loads on buried pipe in a narrow trench condition. Alternatively, the prism load condition can be considered because this type of load is the basis for more conservative flexible pipe theory, which is discussed in the Flexible Pipe Theory section. Simply stated, the prism load is the weight of the column of soil directly over the pipe for the full height of the backfill. This is the maximum load that will be imposed by the soil on a flexible conduit in nearly all cases and is a conservative design approach.

Prism Load: $$W_p = HwB_c \text{ (lb/lin ft)} \tag{4-3}$$

Prism load may also be expressed in terms of soil pressure as follows:

Soil Pressure: $$P = wH \text{ (lb/ft}^2\text{)} \tag{4-4}$$

Where:

P = pressure caused by soil weight at depth H, lb/ft^2
w = unit weight of soil, lb/ft^3
H = depth at which soil pressure is desired, ft

Other Dead or Static Loads

In many cases, the total superimposed load on a PVC pipe is influenced by building foundations, other structure foundations, or other static, long-term loads. These loads may be present at the time that the pipe is installed or may be superimposed on the pipe at some point in time subsequent to the pipe installation. These loads surcharge additional soil pressure onto the buried pipe and can generally be categorized into two types: (1) loads that have confined footprints or areas of influence at the point where the load is transferred to the soil, or (2) loads that have wide areas of influence and normally parallel the pipe. Type 1 loads are usually analyzed as point source loads, while type 2 loads are analyzed as uniformly distributed loads.

The basis for analysis of both types of static loads is the Boussinesq theory, which is mathematically stated as

$$P_A = \frac{WZ^3}{96\pi R^5} \tag{4-5}$$

Where:

P_A = soil pressure at point A, psi
W = superimposed load, lb
Z = vertical distance from the point of the load to the top of the pipe, ft
R = $\sqrt{X^2 + Y^2 + Z^2}$, straight line distance from the point where the load is applied to point A on the top of the pipe; X and Y are measured horizontally at 90° to each other. (Note: Y is usually measured along the horizontal axis of the pipe, and X is usually measured perpendicular to the horizontal axis of the pipe.)

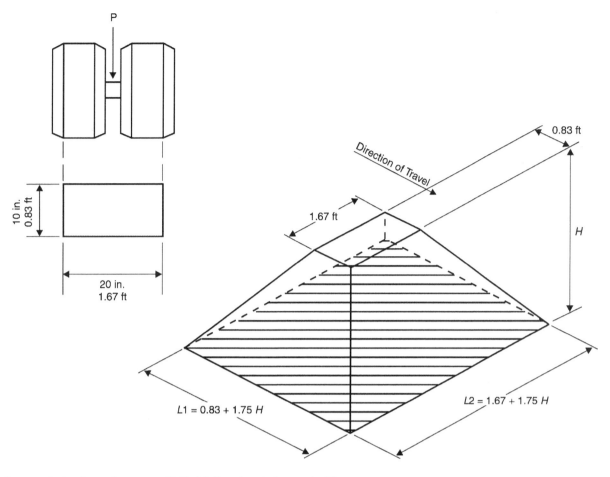

Figure 4-1 Distribution of HS-20 live load through fill

Live Loads

Underground PVC pipe is also subjected to live loads from traffic running over highways, railways, or airport runways, and from other superimposed live loads applied to the surface and transmitted through the soil.

The calculation below assumes a four-lane road with an AASHTO HS-20 truck centered in each 12-ft (3.7 m) wide lane. The pipe may be perpendicular or parallel to the direction of truck travel, or any intermediate position. Other design live loads can be specified as required by project needs and local practice.

1. Compute $L1$, load width (ft) parallel to direction of travel, see Figure 4-1.

$$L1 = 0.83 + 1.75\,H$$

2. Compute $L2$, load width (ft) perpendicular to direction of travel, see Figure 4-1.

$$2' < H < 2.48' \quad L2 = 1.67 + 1.75\,H$$

$$H \geq 2.48' \quad L2 = (43.67 + 1.75\,H)/8$$

3. Compute W_L:

$$W_L = P\,(I_f)/[144(L1)(L2)]$$

Table 4-1 HS-20 and Cooper's E-80 live loads

HS-20 Live Loads				Cooper's E-80 Live Loads			
Depth		W_L		Depth		W_L	
(ft)	(m)	(psi)	(kPa)	(ft)	(m)	(psi)	(kPa)
2.0	0.6	6.0	41.4	4.0	1.2	14.1	97.3
2.5	0.8	3.9	26.9	5.0	1.5	12.2	84.2
3.0	0.9	3.3	22.8	6.0	1.8	10.5	72.5
3.5	1.1	2.6	17.9	8.0	2.4	7.7	53.1
4.0	1.2	2.2	15.2	10.0	3.0	5.7	39.3
6.0	1.8	1.5	10.3	12.0	3.7	4.6	31.7
9.0	2.7	1.0	6.9	14.0	4.3	3.7	25.5
10.0	3.0	0.8	5.5	16.0	4.9	3.0	20.7
12.0	3.7	0.6	4.1	18.0	5.5	2.6	17.9
16.0	4.9	0.5	3.4	20.0	6.1	2.2	15.2
20.0	6.1	0.4	2.8	25.0	7.6	1.5	10.3
27.0	8.2	0.2	1.4	30.0	9.2	1.1	7.6
40.0	12.2	0.1	0.7	35.0	10.7	0.8	5.5
				40.0	12.2	0.6	4.1

Note: Cooper E-80 design loading consisting of four 80,000-lb axles spaced 5 ft c/c. Locomotive load is assumed to be uniformly distributed over an area 8 ft × 20 ft. Weight of track structure is assumed to be 200 lb/lin ft. Impact is included. Height of fill is measured from top of pipe to bottom of ties.

Where:

W_L = live load on pipe, psi
P = 16,000 lb (HS-20 wheel load)
I_f = impact factor
   = 1.1 for 2 ft < H < 3 ft
   = 1.0 for $H \geq 3$ ft

This computation is independent of pipe diameter and results in live loads tabulated in Table 4-1.

FLEXIBLE PIPE THEORY

A flexible pipe may be defined as a conduit that will deflect at least 2 percent without any sign of structural distress, such as injurious cracking. Although this definition is arbitrary, it is widely used.

A flexible pipe derives its soil-load carrying capacity from its flexibility. Under soil load the pipe tends to deflect, thereby developing passive soil support at the sides of the pipe. At the same time, the ring deflection relieves the pipe of the major portion of the vertical soil load, which is then carried by the surrounding soil through the mechanism of an arching action over the pipe. The effective strength of the pipe soil system is quite high.

In flat-plate or three-edge loading, a rigid pipe will support more than a flexible pipe. However, this comparison is misleading if it is used to compare the in-soil capacity of rigid pipe to that of flexible pipe. Flat-plate or three-edge loading is an appropriate measure of load-bearing strength for rigid pipes but not for flexible pipes.

The inherent strength of flexible pipe is called pipe stiffness. It is measured, according to ASTM D2412, *Standard Test Method for External Loading Properties of*

Plastic Pipe by Parallel-Plate Loading, at an arbitrary datum of 5 percent deflection. Pipe stiffness is defined as follows:

$$PS = F/(\Delta Y) \geq \frac{EI}{0.149r^3} = \frac{6.71EI}{r^3} = \frac{6.71Et^3}{12r^3} = 0.559E\left(\frac{t}{r}\right)^3 \tag{4-6}$$

Where:

- PS = pipe stiffness, lbf/in./in.
- F = force, lb/lin in.
- ΔY = vertical deflection, in.
- E = modulus of elasticity, psi
- I = moment of inertia of the wall cross section per unit length of pipe $\left(\frac{t^3}{12} \times 1\right)$, in.^{4}/ lin in. = in.^{3}
- r = mean radius of pipe, in.
- t = wall thickness, in.

For PVC pipe with outside diameter controlled dimensions, Eq 4-6 can be simplified further.

$$DR = \text{dimension ratio} = \frac{D_o}{t}$$

Where: D_o = outside diameter, in.

NOTE: $D_o = 2r + t$ and therefore $r = \frac{D_o - t}{2}$

Substituting into Eq 4-6

$$PS = 0.559E\left(\frac{2t}{D_o - t}\right)^3$$

Simplifying

$$PS = 0.559E \frac{8}{\left(\frac{D_o}{t} - 1\right)^3}$$

Substituting DR for

$$\frac{D_o}{t}$$

and simplifying further

$$PS = 4.47\frac{E}{(DR - 1)^3} \tag{4-7}$$

The resulting *PS* values for various dimension ratios of AWWA C900 and C905 PVC pipe are shown in Table 4-2.

The manner in which flexible pipe performance differs from rigid pipe performance can be understood by visualizing pipe response to applied earth load. In a rigid pipe system, the applied earth load must be carried totally by the inherent strength of the unyielding, rigid pipe, because the soil at the sides of the pipe tends to compress and deform away from the load. In a flexible pipe system, the applied earth load is largely carried by the earth at the sides of the pipe, because the flexible pipe deflects away from the load. That portion of the load carried by the flexible pipe, assumed as a vertical vector of force, is transferred principally through the deflection mechanism into approximately horizontal force vectors assumed by the compressed soil at the sides of the pipe.

Table 4-2 PVC pipe stiffness

DR	Stiffness for Min. E = 400,000 psi (2,758 MPa)	
	lbf/in./in.	(MPa)
51	14	(0.10)
41	28	(0.19)
32.5	57	(0.39)
26	114	(0.79)
25	129	(0.89)
21	224	(1.54)
18	364	(2.51)
14	815	(5.62)

Pressure Class	Stiffness for Min. E = 465,000 psi (3,200 MPa)	
	lbf/in./in.	(MPa)
100	29	(0.20)
150	85	(0.58)
200	201	(1.36)

Because of deflection, the distribution of earth load is carried principally by the surrounding soil envelope and to a lesser extent by the flexible pipe. The strength provided by buried flexible pipe is derived through deflection from the combined strength provided by the pipe-soil system. It should be noted that, in designing water distribution and transmission systems using PVC pressure pipe (AWWA C900, C905, and C909), control or limitation of deflection is usually not a critical design parameter because pipe stiffness values are relatively high and operating internal pressures are usually sufficient to induce rerounding of the pipe.

Spangler's Iowa Deflection Formula

M.G. Spangler, a former student of Anson Marston, observed that the theory of loads on buried rigid pipe was not adequate for flexible pipe design. Spangler noted that flexible pipe may provide little inherent strength in comparison to rigid pipes; yet, when flexible pipe is buried, a significant ability to support vertical loads is derived from the passive pressures induced as the sides of the pipe move outward against the earth. This characteristic, plus the idea that the pipe deflection may also be a basis for design, is reflected in Spangler's Iowa Deflection formula, published in 1941.

The Iowa Deflection formula, as developed by Spangler, was modified in accordance with research and investigation conducted by Dr. R.K. Watkins (Spangler's graduate student) in 1955. The Modified Iowa formula (Eq 4-8) is considered an acceptable approach to theoretical calculation of PVC pipe deflection.

$$\Delta X = D_L K_x \left(\frac{W_c}{12}\right) \frac{r^3}{EI + 0.061 E' r^3} \tag{4-8}$$

Where:

ΔX = horizontal deflection or change in diameter, in.
D_L = deflection lag factor
K_x = bedding constant
W_c = Marston's load per unit length of pipe, lb/lin ft.

E = modulus of elasticity of the pipe material, psi

I = moment of inertia of the pipe wall per unit length, in.^{4}/lin in. = in.^{3}

E' = modulus of soil reaction, psi

r = mean radius of the pipe, in.

Conceptually, the Modified Iowa formula may be rewritten as follows:

$$\text{Deflection} = \text{Constant} \frac{\text{Load}}{\text{Pipe Stiffness} + \text{Soil Stiffness}}$$

Under most soil conditions, PVC pressure pipe tends to deflect minimally into an elliptical shape. The horizontal and vertical deflections may be considered equal in the range of deflections (Δ) allowed in the AWWA PVC pipe standards. Because most PVC pipe is described by either pipe stiffness $F/\Delta Y$ or the ratio (DR) of outside diameter to thickness, the Modified Iowa Deflection formula (Eq 4-8) can be transposed and rewritten as follows:

$$\%\frac{\Delta Y}{D} = \frac{(D_L W_p + W_L)K_x(100)}{0.149\frac{F}{\Delta Y} + 0.061 E'} \tag{4-9}$$

Substituting for pipe stiffness as shown in Eq 4-7 and simplifying further:

$$\%\frac{\Delta Y}{D} = \frac{(D_L W_p + W_L)K_x(100)}{[2E/\ 3(DR-1)^3] + 0.061 E'} \tag{4-10}$$

NOTE: As Eqs 4-9 and 4-10 show, the deflection lag factor, D_L, is not applied to the live load, W_L.

Where:

W_L = live load on the pipe, psi

W_p = prism load (soil pressure), psi

$\dfrac{\Delta Y}{D}$ = percent deflection

Deflection Lag

Unless the operating internal pipe pressure equals or exceeds the external load, a buried flexible pipe will continue to deflect after the full external load is realized. The additional deflection is limited and is a function of soil density in the pipe zone. As soil density at the sides of the pipe increases, the total deflection in response to load decreases. After the trench load reaches a maximum, the pipe-soil system continues to deflect only as long as the soil is in the process of consolidation. Once the soil has reached the density required to support the load, the pipe will not deflect further.

A full load on any buried pipe is not reached immediately after installation unless the final backfill is compacted to a high density. For flexible pipe, the long-term load will not exceed the prism load. Therefore, for design, the prism load can be used to effectively compensate for the increased trench consolidation load with time.

Deflection lag factor D_L. The deflection lag factor, D_L, converts the immediate deflection of the pipe to the deflection of the pipe after many years. The primary cause of increasing pipe deflection with time is the increase in overburden load as soil "arching" is gradually lost. The vast majority of this phenomenon occurs during the first few months of burial or up to seven years, depending on the frequency of wetting and drying cycles. Secondary causes of increasing pipe deflection with time are the time-related consolidation of the pipe zone embedment and the creep of the native soil at the sides of the pipe. These causes are generally of much less significance than

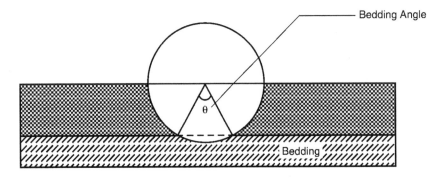

Figure 4-2 Bedding angle

Table 4-3 Bedding constant values

Values of Bedding Constant, K_x	
Bedding Angle (*degrees*)	K_x
0	0.110
30	0.108
45	0.105
60	0.102
90	0.096
120	0.090
180	0.083

increasing load and may not contribute to the deflection for pipes buried in relatively stiff native soils with dense granular pipe zone surroundings. For long-term deflection prediction, a D_L value greater than 1.00 is appropriate when using the Marston Flexible Pipe Load (Eq 4-2) for the dead load.

Alternatively, when the conservative prism load, W_P (discussed below), is used for the dead load, a D_L value equal to 1.00 is appropriate.

Bedding coefficient K_x. The bedding coefficient, K_x, reflects the degree of support provided by the soil at the bottom of the pipe and over which the bottom reaction is distributed. Assuming an inconsistent haunch achievement (typical direct bury condition), a K_x value of 0.1 should be used. For uniform shaped bottom support, a K_x value of 0.083 is appropriate.

The bedding angle (θ) is shown in Figure 4-2. Bedding constant values are shown in Table 4-3.

Prism load on the pipe W_p. As previously discussed, the vertical soil load on the pipe may be considered as the weight of the rectangular prism of soil directly above the pipe. The soil prism would have a height equal to the depth of earth cover and a width equal to the pipe outside diameter.

$$W_p = \frac{\gamma_s H}{144} \qquad (4\text{-}11)$$

Where:

W_p = prism load, lb/in.^{2}

γ_s = unit weight of overburden, lb/ft^3

H = burial depth of top of pipe, ft

Table 4-4 Values for the soil support combining factor, S_c

E'_n/E'_b	B_d/D					
	1.5	2.0	2.5	3	4	5
0.1	0.15	0.30	0.60	0.80	0.90	1.00
0.2	0.30	0.45	0.70	0.85	0.92	1.00
0.4	0.50	0.60	0.80	0.90	0.95	1.00
0.6	0.70	0.80	0.90	0.95	1.00	1.00
0.8	0.85	0.90	0.95	0.98	1.00	1.00
1.0	1.00	1.00	1.00	1.00	1.00	1.00
1.5	1.30	1.15	1.10	1.05	1.00	1.00
2.0	1.50	1.30	1.15	1.10	1.05	1.00
3.0	1.75	1.45	1.30	1.20	1.08	1.00
≥5.0	2.00	1.60	1.40	1.25	1.10	1.00

Note: In-between values of S_c may be determined by straight-line interpolation from adjacent values.

Modulus of soil reaction E'. The vertical loads on a flexible pipe cause a decrease in the vertical diameter and an increase in the horizontal diameter. The horizontal movement develops a passive soil resistance that acts to help support the pipe. The passive soil resistance varies depending on the soil type and the degree of compaction of the pipe-zone backfill material, the native soil characteristics (Table 4-6), the cover depth, and the trench width. To determine E' for a buried pipe, separate E' values for the native soil, E'_n, and the pipe backfill surround, E'_b, must be determined and then combined using Eq 4-12.

$$E' = S_c E'_b \qquad (4\text{-}12)$$

Where:

- E' = composite modulus of soil reaction, psi (to be used in Eq 4-9 and 4-10)
- S_c = soil support combining factor from Table 4-4, dimensionless
- E'_b = modulus of soil reaction of the pipe-zone embedment from Tables 4-5 and 4-6, psi

To use Table 4-4 for S_c values, the following values must be determined:

- E'_n = modulus of soil reaction of the native soil at pipe elevation from Table 4-7, psi
- B_d = trench width at pipe springline, in.

Design Example No. 1

What will be the deflection of a DR 18 PVC pipe buried in a flat-bottom trench? If a Type 2 embedment as shown in AWWA C605 is assumed, then modulus of soil reaction (E'_b) for the embedment is 200 psi. Also assume the following:

Pipe dia. (D) = 12 in.

Trench width at pipe springline (B_d) = 18 in.

Cover depth (H) = 10 ft.

Bedding coefficient (K_x) = 0.110

Modulus of soil reaction (E'_n) for native soil = 2,000 psi

Native soil specific weight (γ_s) = 120 lb/ft³

Lag factor (D_L) = 1.0

Table 4-5 Values for the modulus of soil reaction E'_b for the pipe-zone embedment, psi (MPa)

Soil Stiffness Category	Soil Type-Primary Pipe-Zone Backfill Material (Unified Classification System)*	Dumped	Slight <85% Proctor <40% Relative Density	Moderate 85-95% Proctor 40-70% Relative Density	High >95% Proctor >70% Relative Density
SC5	Fine-grained soil with medium to high plasticity: (CH, MH, OL, OH, PT, or borderline soils CH/MH, and so forth)	Soils in this category require special engineering analysis to determine required density, moisture content, and compactive effort.	Soils in this category require special engineering analysis to determine required density, moisture content, and compactive effort.	Soils in this category require special engineering analysis to determine required density, moisture content, and compactive effort.	Soils in this category require special engineering analysis to determine required density, moisture content, and compactive effort.
SC4	Fine-grained soils with medium to no plasticity: (CL, ML, ML-CL, or borderline soil ML/MH, and so forth) with <30% coarse-grained particles	50 (0.34)	200 (1.4)	400 (2.8)	1,000 (6.9)
SC3	Fine-grained soil with medium to no plasticity: (CL, ML, ML-CL, or borderline soil CL-CH, and so forth) with ≥30% coarse-grained particles	100 (0.69)	400 (2.8)	1,000 (6.9)	2,000 (13.8)
SC3	Course-grained soil with fines (GM, GC, SM, SC, GM-GC, GC-SC, and so forth) containing more than 12% fines	100 (0.69)	400 (2.8)	1,000 (6.9)	2,000 (13.8)
SC2	Course-grained soils with little or no fines (GW, GP, SW, SP, GW-GC, SP-SM, and so forth) containing 12% fines and less	200 (1.4)	1,000 (6.9)	2,000 (13.8)	3,000 (20.7)
SC1	Crushed rock and GP with ≤15% sand, max. 25% passing the ⅜-in. sieve, and max. 5% fines.	1,000 (6.9)	3,000 (20.7)	3,000 (20.7)	3,000 (20.7)

* ASTM Classification D2487 (see Table 4-6).
Percent proctor density per ASTM D698 and relative density per ASTM D4253 and D4254.
Values for E'_b for in-between soils or borderline proctor densities may be interpolated.

Table 4-6 Soil classification chart (ASTM D2487)

	Criteria for Assigning Group Symbols and Group Names Using Laboratory Tests[a]			Soil Classification	
				Group Symbol	Group Name[b]
Coarse-grained soils More than 50% retained on No. 200 sieve	More than 50% of coarse fraction retained on No. 4 sieve	Gravels	Clean gravels Less than 5% fines[e]		
			$Cu \geq 4$ and $1 \leq Cc \leq 3$[c]	GW	Well-graded gravel[d]
			$Cu < 4$ and/or $1 > Cc > 3$[c]	GP	Poorly graded gravel[d]
			Gravels with fines More than 12% fines[e]		
			Fines classify as ML or MH	GM	Silty gravel[d,f,g]
			Fines classify as CL or CH	GC	Clayey gravel[d,f,g]
	50% or more of coarse fraction passes No. 4 sieve	Sands	Clean sands Less than 5% fines[i]		
			$Cu \geq 6$ and $1 \leq Cc \leq 3$[c]	SW	Well-graded sand[h]
			$Cu < 6$ and/or $1 > Cc > 3$[c]	SP	Poorly graded sand[h]
			Sands with fines More than 12% fines[i]		
			Fines classify as ML or MH	SM	Silty sand[f,g,h]
			Fines classify as CL or CH	SC	Clayey sand[f,g,h]
Fine-grained soils 50% or more passes the No. 200 sieve	Silts and Clays Liquid limit less than 50	Inorganic	PI > 7 and plots on or above "A" line[j]	CL	Lean clay[k,l,m]
			PI < 4 or plots below "A" line[j]	ML	Silt[k,l,m]
		Organic	Liquid limit—oven dried < 0.75 Liquid limit—not dried	OL	Organic clay[k,l,m,n] Organic silt[k,l,m,o]
	Silts and Clays Liquid limit 50 or more	Inorganic	PI plots on or above "A" line	CH	Fat clay[k,l,m]
			PI plots below "A" line	MH	Elastic silt[k,l,m]
		Organic	Liquid limit—oven dried < 0.75 Liquid limit—not dried	OH	Organic clay[k,l,m,p] Organic silt[k,l,m,q]
Highly organic soils	Primarily organic matter, dark in color, and organic odor			PT	Peat

Source: Reprinted with permission, ASTM D2487-00, copyright American Society of Testing and Materials, West Conshohocken, Pa.

a Based on the material passing the 3-in. (75-mm) sieve.
b If field sample contained cobbles and/or boulders, add "with cobbles and/or boulders" to group name.
c $Cu = D_{60}/D_{10}$
   $Cc = \dfrac{(D_{30})^2}{D_{10} \times D_{60}}$
d If soil contains ≥15% sand, add "with sand" to group name.
e Gravels with 5 to 12% fines require dual symbols:
   GW-GM well-graded gravel with silt
   GW-GC well-graded gravel with clay
   GP-GM poorly graded gravel with silt
   GP-GC poorly graded gravel with clay
f If fines classify as CL-ML, use dual symbol GC-GM or SC-SM.
g If fines are organic, add "with organic fines" to group name.
h If soil contains ≥15% gravel, add "with gravel" to group name.
i Sands with 5 to 12% fines require dual symbols:
   SW-SM well-graded sand with silt
   SW-SC well-graded sand with clay
   SP-SM poorly graded sand with silt
   SP-SC poorly graded sand with clay
j If the Atterberg limits (liquid limit and plasticity index) plot in hatched area on plasticity chart, soil is a CL-ML, silty clay.
k If soil contains 15 to 29% plus No. 200, add "with sand" or "with gravel," whichever is predominant.
l If soil contains ≥30% plus No. 200, predominantly sand, add "sandy" to group name.
m If soil contains ≥30% plus No. 200, predominantly gravel, add "gravelly" to group name.
n PI ≥ 4 and plots on or above "A" line (Fig. 4-3).
o PI ≤4 or plots below "A" line (Fig. 4-3).
p PI plots on or above "A" line (Fig. 4-3).

32   PVC PIPE—DESIGN AND INSTALLATION

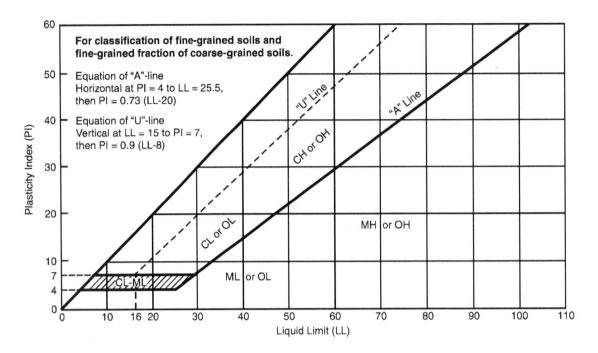

Source: Reprinted with permission, ASTM D2487-00, copyright American Society of Testing and Materials, West Conshohocken, Pa.

Figure 4-3   Plasticity chart

Table 4-7   Values for the modulus of soil reaction, E'_n, for the native soil at pipe-zone elevation

Native in situ soils					
Granular		Cohesive		
Blows/ft*	Description	q_u (Tsf)	Description	E'_n (psi)
>0–1	very, very loose	>0–0.125	very, very soft	50
1–2	very loose	0.125–0.25	very soft	200
2–4	very loose	0.25–0.50	soft	700
4–8	loose	0.50–1.0	medium	1,500
8–15	slightly compact	1.0–2.0	stiff	3,000
15–30	compact	2.0–4.0	very stiff	5,000
30–50	dense	4.0–6.0	hard	10,000
>50	very dense	>6.0	very hard	20,000
Rock				≥50,000

Source: Pipeline Installation by Amster Howard, Lakewood, Colorado, 1996.

* Standard penetration test per ASTM D1586.

Notes: E' Special cases:

 1) Geotextiles—When a geotextile pipe zone wrap is used, E'_n values for poor soils can be greater than shown in Table 4-7.
 2) Solid sheeting—When permanent solid sheeting designed to last the life of the pipeline is used in the pipe zone, E' shall be based solely on E'_b.
 3) Cement-stabilized sand—When cement-stabilized sand is used as the pipe-zone surround, initial deflections shall be based on a sand installation and the long term E'_b = 25,000 psi. (Typical mix ratio is one sack of cement per ton or 1.5 sacks of cement per cubic yard of mix.)
 4) For embankment installation $E'_b = E'_n = E'$.

No internal operating pressure

$$\frac{E'_n}{E'_b} = \frac{2{,}000}{200} = 10$$

$$\frac{B_d}{D} = \frac{18}{12} = 1.5$$

From Table 4-4, $S_c = 2.0$

$$E' = S_c E'_b = 2(200) = 400$$

$$W_p = \frac{\gamma_s H}{144} = \frac{120 \times 10}{144} = 8.33$$

Modulus of elasticity (E') of PVC material = 400,000 psi

Liveload ($W_L = 0$)

$$\%\frac{\Delta Y}{D} = \frac{(D_L W_P + W_L)K_x(100)}{(2E'/\ 3(DR-1)^3) + 0.061 E'}$$

$$\%\frac{\Delta Y}{D} = \frac{(1.0 \times 8.33 + 0)0.110(100)}{(2 \times 400{,}000/\ 3(18-1)^3) + 0.061(400)}$$

$$\%\frac{\Delta Y}{D} = \frac{91.63}{54.28 + 24.40} = 1.16\%$$

LONGITUDINAL BENDING

The ability of PVC pipe to withstand some longitudinal bending is considered a significant advantage in buried applications. Some longitudinal bending may be effected deliberately during PVC pipe installation to make changes in alignment to avoid obstructions. Bending also may occur in response to various unplanned conditions or unforeseen changes in conditions in the pipe-soil system, such as the following:

- Differential settlement of a large valve or structure to which the pipe is rigidly connected
- Ground movement associated with earthquakes
- Uneven settlement of the pipe bedding
- Ground movement associated with excessive surface or traffic loading
- Ground movement associated with tidal or groundwater conditions
- Seasonal variation in soil conditions caused by changes in moisture content (limited to expansive or organic soils)
- Improper installation procedures: for example, nonuniform foundation, unstable bedding, or inadequate embedment consolidation

Through longitudinal bending, PVC pipe provides the ability to deform or bend and move away from external pressure concentrations. The use of flexible joints also enhances a pipe's ability to yield to these forces, thereby reducing risk of damage or failure. Proper engineering design and installation will limit longitudinal bending of PVC pipe within acceptable criteria.

The following discussion regarding allowable bending equally applies to PVCO pipe.

Allowable Longitudinal Bending

When installing PVC pipe, some changes in alignment of the pipe may be accomplished without the use of elbows, sweeps, or other direction-change fittings. Controlled longitudinal bending of the pipe within acceptable limits can be accommodated by PVC pipe itself. Longitudinal deviation of the pipeline can be accommodated through either joint offset or axial flexure (curving) of the pipe. *Curving of PVC pipe may increase the possibility of failures during tapping under pressure. Whenever practicable, taps should be made on the inside radius of longitudinally bent pipe.*

Permissible joint offset may be significant when gasketed joints that are designed for such a purpose are provided on the PVC pipe. Depending on pipe size and joint design, the offset per joint for gasketed PVC pipe joints in the unstressed condition varies from approximately ⅓° to 5°. Available joint offset is dependent upon insertion depth of the spigot. Assembly should be made only to the mark provided. "Homed" or over-inserted assemblies allow for no flexibility. Some manufacturers encourage joint offset only for accommodation of laying water pipe not to grade, with fittings to be used for known changes in direction. Joint offset limits should be obtained from the manufacturer for joints that are to be stressed to the permissible limit without leakage.

Mathematical relationships for the longitudinal bending of pressurized tubes have been derived by Reissner. These relationships compare favorably to those of Timoshenko and others. One critical limit to bending of PVC pipe is long-term flexural stress. Axial bending also causes a very small amount of ovalization or diametric deflection of the pipe.

AWWA PVC pipe has short-term strengths of 7,000–8,000 psi (48.26–55.16 MPa) in tension and 11,000–15,000 psi (75.84–103.42 MPa) in flexure. The long-term strength of PVC pipe in either tension, compression, or flexure can conservatively be assumed to equal the hydrostatic design basis *HDB* of 4,000 psi (27.58 MPa). Applying a 2.5:1 safety factor results in an allowable long-term tensile or flexural stress equal to the recommended hydrostatic stress *S* of 1,600 psi (11.03 MPa) for AWWA C900 PVC pipe at 73.4° F (23° C). The 2:1 safety factor used for AWWA C905 PVC pipe similarly yields a hydrostatic stress of 2,000 psi (13.79 MPa). However, the bending of PVC pipe barrels larger than 12 in. (300 mm) is usually not recommended because of the forces required. Either the 1,600 psi (11.03 MPa) or 2,000 psi (13.79 MPa) figure may be used for the allowable long-term flexural stress in gasketed joint pipe that is free of longitudinal stress from internal pressure longitudinal thrust. However, when the joints are restrained as in solvent cementing, without snaking the pipe in the trench, then the end thrust from internal pressure imposes a longitudinal tensile stress equal to one-half of the hoop stress.

The equation for allowable bending stress S_b is

$$S_b = (HDB - S_t)\frac{T'}{SF} \qquad (4\text{-}13)$$

Where:

HDB = hydrostatic design basis of PVC pipe, psi [4,000 psi (27.58 MPa) for AWWA C900 and C905]

S_t = HDB/2 = tensile stress from longitudinal thrust, psi

T' = temperature rating factor (see Table 5-1)

SF = safety factor (2.5 or 2.0 as applicable)

NOTE: The longitudinal stress from thermal expansion and contraction can be ignored in buried gasketed joint piping because of relaxation of the soil restraint over the length between joints. When determining bending limits, longitudinal thermal stresses should be considered in restrained pipes, such as lines with solvent-cemented joints and restrained and supported piping.

Using Eq 4-13, the maximum allowable bending stress (S_b) for restrained joint AWWA PVC pipe is calculated as follows.

$$(S_b) \text{ at } 73.4°F \text{ } (23°C)$$

$$\text{Pressure Class Pipe (AWWA C900)} = \left[4{,}000 - \frac{4{,}000}{2}\right] \frac{1.0}{2.5} = 800 \text{ psi (5.52 MPa)}$$

$$\text{Pressure Class Pipe (AWWA C905)} = \left[4{,}000 - \frac{4{,}000}{2}\right] \frac{1.0}{2.0} = 1{,}000 \text{ psi (6.89 MPa)}$$

NOTE: It is emphasized that construction machinery *must not be used* to bend PVC.

The mathematical relationship between stress and moment induced by longitudinal bending of pipes is

$$M = \frac{S_b I}{c} \tag{4-14}$$

Where:

M = bending moment, in.-lb
S_b = allowable bending stress, psi
c = $D_o/2$ = distance from extreme fiber to neutral axis, in.
I = moment of inertia, in.^{4}

$$\left(\frac{\pi}{64}\right)(D_o^4 - D_i^4) = 0.049087(D_o^4 - D_i^4) \tag{4-15}$$

Where:

D_o = average outside diameter, in.
D_i = average inside diameter, in.
   = $D_o - 2t_{avg}$, where:
t_{avg} = average wall thickness, in.
t_{avg} = t_{min} + 6% of t_{min}
t_{min} = minimum wall thickness, in.

Assuming that the bent length of pipe conforms to a circular arc after backfilling and installation (Figure 4-4), the minimum radius R_b of the bending circle can be found by Timoshenko's equation:

$$R_b = \frac{EI}{M} \tag{4-16}$$

Combining Eq 4-14 and 4-16 gives

$$R_b = \frac{ED_o}{2S_b} \tag{4-17}$$

The central angle β subtended by the length of pipe is

$$\beta = \frac{360L}{2\pi R_b} = \frac{57.30L}{R_b} \tag{4-18}$$

Where:

R_b = minimum bending radius, in.
E = modulus of tensile elasticity, psi
M = bending moment, in.-lb
S_b = allowable bending stress, psi
L = pipe length, in.

36 PVC PIPE—DESIGN AND INSTALLATION

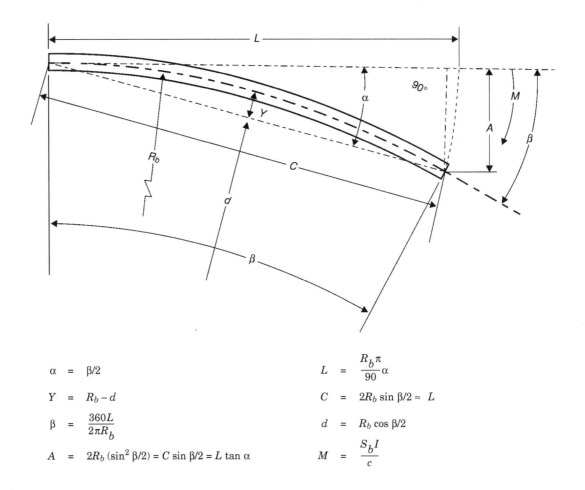

$\alpha = \beta/2$

$Y = R_b - d$

$\beta = \dfrac{360L}{2\pi R_b}$

$A = 2R_b (\sin^2 \beta/2) = C \sin \beta/2 = L \tan \alpha$

$L = \dfrac{R_b \pi}{90} \alpha$

$C = 2R_b \sin \beta/2 \approx L$

$d = R_b \cos \beta/2$

$M = \dfrac{S_b I}{c}$

Figure 4-4 PVC pipe longitudinal bending

L and R_b are both in the same units, and the angle of lateral offset (α) of the curved pipe from a tangent to the circle is

$$\alpha = \frac{\beta}{2}, \text{degrees} \tag{4-19}$$

The offset A at the end of the pipe from the tangent to the circle can then be calculated as

$$A = 2R_b (\sin^2 \beta/2) = 2R_b (\sin \alpha)^2 \tag{4-20}$$

Where:

A = offset at the end of the pipe from the tangent to the circle, in.

Assuming that during installation the pipe is temporarily fixed at one end and acts as a cantilevered beam, then the lateral force required at the free end to achieve the offset A may be determined by the following equation:

$$P = \frac{3EIA}{L^3} \tag{4-21}$$

Where:

P = lateral offset force, lb

E = modulus of tensile elasticity, psi

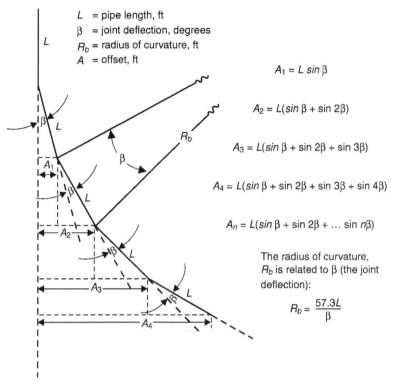

Figure 4-5 PVC pipe joint offset

I = moment of inertia, in.^{4}
A = offset at free end, in.
L = pipe length, in.

The mathematical relationship between the bending offset angle α, the offset A, the lateral offset force P, and the minimum bending radius R_b is shown in Figure 4-4. The bending limits calculated by the equations listed above are without allowance for joint offset and without consideration of the stresses imposed upon the joint. Because of the characteristics of a particular joint design, it is possible that a manufacturer's recommended bending radius may be greater or lesser than those tabulated. Several manufacturers currently recommend a bending radius for PVC pipe of about $R_b = 300\,D_o$.

Joint Offset

When a desired change of direction in a PVC pipeline exceeds the permissible joint offset angle β for a given length of pipe, the total offset required can be distributed through several pipe lengths (Figure 4-5). Calculation of required distribution of total offset in PVC pipe is demonstrated in the following example.

- Pipeline using AWWA C900 8-in. PVC DR 18 pipe in 20-ft lengths.
- Desired change of direction is 10°.
- End offset of one 20-ft length = A1 (Figure 4-5).

    $\beta = 3.0°$ per pipe length, maximum end offset

    (Four lengths of 8-in. × 20-ft pipe at 2.5° are required.)

38 PVC PIPE—DESIGN AND INSTALLATION

Table 4-8 Longitudinal bending stress and strain in PVC pipe at 73.4°F (23°C)

Bending Radius Ratio R_b/D_o	Elastic Modulus E		Bending Strain ε	Bending Stress S_b	
	psi	(MPa)	in./in.	psi	(MPa)
25	400,000	(2,758)	0.0200	8,000	(55.16)
50	400,000	(2,758)	0.0100	4,000	(27.58)
100	400,000	(2,758)	0.0050	2,000	(13.79)
200	400,000	(2,758)	0.0025	1,000	(6.89)
250	400,000	(2,758)	0.0020	800	(5.52)
300	400,000	(2,758)	0.0017	667	(4.60)
500	400,000	(2,758)	0.0010	400	(2.76)

Note: These values also apply to PVCO pipe.

Resultant total offset for the pipeline over four pipe lengths:

$$A_4 = 20[\sin 2.5° + \sin(2 \times 2.5°) + \sin(3 \times 2.5°) + \sin(4 \times 2.5°)]$$
$$= 20 (0.0436 + 0.0872 + 0.1305 + 0.1736)$$
$$= 20 (0.4349)$$
$$= 8.7 \text{ ft}$$

Performance Limits in Longitudinal Bending

Bending strain. Longitudinal bending strain ε and longitudinal bending stress S_b for PVC pipe at different bending radius ratios are tabulated in Table 4-8 using Eq 4-22:

$$\varepsilon = S_b/E = D_o/2R_b \qquad (4\text{-}22)$$

Bending ovalization (diametric or ring deflection). As a thin tube is bent longitudinally, it will change into an approximate elliptical shape. This effect has been ignored as insignificant in previous calculations on longitudinal bending. Ring deflection is usually expressed as

$$\text{Deflection} = \delta = \frac{\Delta Y}{D} \qquad (4\text{-}23)$$

$$\% \text{ Deflection} = 100\delta = 100 \frac{\Delta Y}{D} \qquad (4\text{-}24)$$

Where:
ΔY = reduction in diameter, in.
D = diameter, in.

The mathematical relationships for thin pressurized tubes between ring deflection and axial bending, as derived by E. Reissner, are as follows:

$$\delta = \frac{\Delta Y}{D} = -(A_1\alpha^2)\left[\left(\frac{2}{3}\right) + \frac{71 + 4\lambda}{135 + 9\lambda}(A_1\alpha^2)\right] \qquad (4\text{-}25)$$

with λ and $(A_1\alpha^2)$ defined as:

$$\lambda = \frac{12(1-u^2)PD_m^3}{8Et^3} \qquad (4\text{-}26)$$

$$(A_1\alpha^2) = \frac{1}{6}\left[\frac{18(1-u^2)}{12+4\lambda}\right]\frac{D_m^4}{R^2 t^2} \qquad (4\text{-}27)$$

Where:

D_m = mean pipe diameter, in
u = Poisson's ratio (0.38 for PVC)
P = internal pipe pressure, psig
E = modulus of elasticity, psi
t = pipe thickness, in. (use $t_{\text{nom}} = 1.06 \times t$)
R = bending radius of pipe, in.

Example: Calculate the percent ring deflection that results from bending a 4-in., DR 14 PVC pressure pipe to its minimum bending radius of 300 times the diameter, given that the pipe is pressurized to 100 psi:

$$\lambda = \frac{12(1-0.38^2)100(4.800-0.364)^3}{8(400,000)(0.364)^3}$$

$$= \frac{1,200(0.8556)(87.292)}{3,200,000(0.04823)} = 0.581$$

$$(A_1\alpha^2) = \frac{1}{16}\left[\frac{18(1-0.38^2)}{12+4(0.581)}\right]\frac{(4.800-0.364)^4}{(1,200)^2(0.364)^2}$$

$$= \frac{1}{16}\left[\frac{18 \times 0.8556}{12+2.324}\right]\frac{387.2275}{(1,440,000)(0.1325)}$$

$$= 0.000136$$

$$\delta = -(0.000136)\left[\left(\frac{2}{3}\right) + \frac{71+4(0.581)}{135+9(0.581)} \times 0.000136\right]$$

$$= -(0.000136)\left[\frac{2}{3} + 0.0000711\right]$$

$$= -0.0000907$$

$$= -0.009\%$$

A negative value is normal and indicates that the deflection deforms the pipe such that the diameter paralleling the plane of bending is decreased and the diameter perpendicular to the plane of bending is increased.

The preceding example shows that at the recommended maximum bending (minimum bending radius) for 4–12-in. AWWA PVC pressure pipes, a close approximation of deflection can be calculated from the following equation:

$$\delta = \frac{\Delta Y}{D_m} = -\frac{2}{3}(A_1\alpha^2) = -\frac{(1-u^2)(D_m)^4}{16R^2t^2} \quad (4\text{-}28)$$

Analysis of similar examples has shown that the amount of deflection resulting from bending is negligible. Generally, at bending radii of 300 times the diameter, the diametric ring deflection from bending will be less than 0.06 percent.

EXPANSION AND CONTRACTION

All pipe products expand and contract with changes in temperature. Linear expansion and contraction of any unrestrained pipe on the longitudinal axis relates to the coefficient of thermal expansion for the specific material used in manufacturing the product. Variation in pipe length as a result of thermal expansion or contraction depends on the pipe material's coefficient of thermal expansion and the variation in temperature, ΔT. Changes in pipe diameter or wall thickness do not produce a change in rates of thermal expansion or contraction.

Coefficients of Thermal Expansion

Approximate coefficients of thermal expansion for different pipe materials are presented in Table 4-9. Expansion and contraction of pipe in response to change in temperature will vary slightly with changes in the exact composition of the pipe material. However, the coefficients shown in Table 4-9 can be considered reasonably accurate. Table 4-10 displays typical length variation of unrestrained PVC pipe as a result of thermal expansion and contraction. PVC pipe length variation caused by temperature change is shown graphically in Figure 4-6.

A good general rule in design of PVC piping systems is to allow ⅜ in. (10 mm) of length variation for every 100 ft (30 m) of pipe for each 10°F (5.6°C) change in temperature. The coefficient of thermal expansion is the same value for PVCO and PVC pipe material.

PVC pipe with gasketed joints, if properly installed (i.e., with pipe spigot inserted into bell joints up to manufacturer's insertion mark), will accommodate substantial thermal expansion and contraction. If gasketed joints are used, thermal

Table 4-9 Coefficients of thermal expansion

Piping Material	Coefficient $in./in./°F$	Expansion $in./100 ft/10°F$
PVC	3.0×10^{-5}	0.36
HDPE	1.2×10^{-4}	1.44
ABS	5.5×10^{-5}	0.66
Asbestos cement	4.5×10^{-6}	0.05
Aluminum	1.3×10^{-5}	0.16
Cast iron	5.8×10^{-6}	0.07
Ductile iron	5.8×10^{-6}	0.07
Steel	6.5×10^{-6}	0.08
Clay	3.4×10^{-6}	0.04
Concrete	5.5×10^{-6}	0.07
Copper	9.8×10^{-6}	0.12

Table 4-10 Length variation per 10°F (5.6°C) ΔT for PVC pipe

Pipe Length		Length Change	
ft	(m)	in.	(mm)
20	(6.1)	0.072	(1.83)
10	(3.0)	0.036	(0.91)

DESIGN FACTORS 41

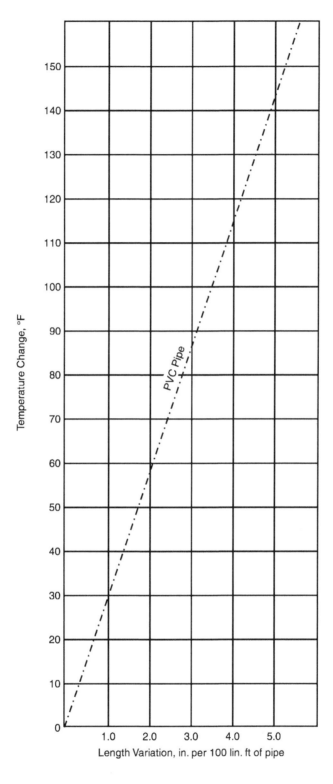

Figure 4-6 Length variation of unrestrained PVC pipe as a result of temperature change

expansion and contraction is not a significant factor in system design within the accepted range of operating temperatures for buried PVC water pipe.

For systems where expansion and contraction cannot be accommodated by gasketed joints, the stresses resulting from extreme temperature changes should be considered using the following equation:

$$S' = EC_T(t_1 - t_o) \qquad (4\text{-}29)$$

Where:

S' = stress, psi
E = modulus of tensile elasticity, psi
C_T = coefficient of thermal expansion, in./in./°F
t_1 = highest temperature, °F
t_o = lowest temperature, °F

Example: Calculate the stress resulting from a temperature change from 120°F to 30°F in a restrained PVC pipe.

S' = $(400{,}000)(3.0 \times 10^{-5})(120 - 30)$
S' = 1,080 psi

This is below the minimum short-term tensile strength of 7,000 psi. The 7,000 psi value may be used as a conservative value for the short-term compressive strength.

THRUST RESTRAINT—GENERAL

Water under pressure exerts thrust forces in a piping system. Internal hydrostatic pressure acts perpendicularly on any plane with a force equal to the pressure times the area of the plane. In straight lengths of a water main, the forces balance each other, assuming that both the pressure and size are equal.

Unbalanced forces, however, are developed at elbows, tees, wyes, reducers, valves, and dead ends. Unbalanced forces are a result of one or more of the following conditions:

- Operating pressure
- Transient pressure
- Velocity head
- Changing direction of flow
- Changing the diameter of the pipe
- Thermal changes
- Dead weight of pipe and liquid

The setting of the design parameters and the selection of the appropriate method of providing thrust restraint must be based on the consideration of present and future conditions. The long-range considerations should include the following:

- The prediction of future activities around the bends that might jeopardize the water main, such as construction of other services, changes in groundwater levels, and changes in ground elevations
- The ability to control and monitor activities around the bends, especially those on easements away from public surveillance
- The risks to the public from major failures both in regard to property damage and public health

As is the case for designing buildings, the design of thrust restraint systems is the responsibility of an engineer with experience in water main construction, as well as geotechnical engineering. All the unbalanced forces at a bend must be identified and the systems designed to restrain these forces at all times in a safe and permanent manner with due consideration as to future predictable and unpredictable conditions. Designs cannot be left to chance or to unqualified persons who rely only on formulae from manuals, magazine articles, or manufacturers' information. The use of the formulae and manufacturers' recommended devices and theory must be thoroughly understood and appraised.

Thrust Blocks

One common method of resisting the unbalanced forces in a water main is a thrust block. Typical thrust blocks are shown in Figure 7-7. Thrust blocks, if properly designed and installed, provide a satisfactory method of resisting the unbalanced forces in both a lateral and vertical direction.

Some of the advantages of thrust blocks are as follows:

- Inexpensive to install
- Easy to install
- Corrosion resistant

Thrust blocks also have disadvantages that must be carefully assessed. Thrust blocks rely on the undisturbed soil behind the thrust block. The reliability of the thrust block may be compromised if digging occurs in the vicinity of the back of the thrust block in the future. Thrust blocks should be shown on the water main system plans and a cautionary note about digging in the area should be on the drawing.

In some cases, the actual size of the block will prohibit its use. This is particularly true in many urban areas. In poor soil conditions, large and heavy thrust blocks may cause settlement of the pipe. Compact fittings have made the construction of thrust blocks more difficult because of the small area of bend available to pour the thrust block against without encasing the joint.

Thrust Block Design Criteria

The design and installation of thrust blocks can be a critical part of a successful water main project. It is important that the soil-bearing capacity be carefully and properly assessed. The design engineer is responsible for selecting the bearing strength of the undisturbed native soil. It is recommended that the bearing strength of the soil be determined by a qualified geotechnical engineer. The present and possible future water table elevation must also be considered when assuming the bearing capacity of the soil.

The major design criteria for thrust blocks are as follows:

- The thrust block bearing surface should be placed against undisturbed soil or compacted backfill.
- The vertical height of the thrust block should be less than half the total depth to the final ground surface.
- The width of the thrust block should be at least two times the height of the thrust block.
- The designer must consider the maximum operating pressure, transient pressure, and test pressure in computing the unbalanced force.

The forces at a horizontal bend are shown in Figure 4-7. The total unbalanced forces at the bend are as follows:

$$T = 2PA \sin \Delta/2 \qquad (4\text{-}30)$$

Where:

T = resultant thrust force, lb
P = internal pressure, psi

44 PVC PIPE—DESIGN AND INSTALLATION

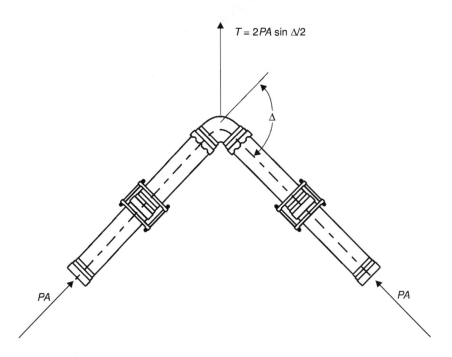

Figure 4-7 Free-body diagram of forces on a pipe bend

A = internal area based on the diameter of the sealing element, in.^{2}

Δ = angle of deflection, degrees

The internal area A must be based on the maximum inside diameter of the sealing element. In the case of joints in a pipeline where the gasket is carried by the bell, the internal area A is based on the pipe outside diameter at the joint.

The required thrust block bearing area can be calculated by the following equation:

$$\text{Area} = \frac{T}{S_b}(SF) \qquad (4\text{-}31)$$

Where:

Area = area of thrust block, ft^2

T = resultant thrust force, lb.

S_b = soil bearing capacity, lb/ft^2

SF = safety factor (suggest 1.5)

The soil allowable bearing capacity should be evaluated by a geotechnical engineer. Some typical values of the soil bearing capacity are presented in Table 4-11.

A passive resistance thrust block design is required if the height of the thrust block is greater than 0.5 times the depth.

The required thrust block area using this method can be calculated as follows:

$$A = \frac{T(SF)}{\gamma H_t N_d + 2C_s \sqrt{N_d}} \qquad (4\text{-}32)$$

Where:

A = area of thrust block, ft^2

T = unbalanced force, lb

SF = safety factor

γ = unit weight of soil, lb/ft^3

DESIGN FACTORS   45

Table 4-11 Estimated bearing strength (undisturbed soil)

Soil Type	lb/ft²
Muck, peat, etc.	0
Soft clay	500
Sand	1,000
Sand and gravel	1,500
Sand and gravel with clay	2,000
Sand and gravel, cemented with clay	4,000
Hard pan	5,000

Source: Buried Pipe Design by Dr. Moser, New York, 1990.

H_t = total depth to bottom of block, ft
N_d = $\tan^2(45° + \phi/2)$
   ϕ = soil internal friction angle (degrees)
C_s = soil cohesion, lb/ft²

For vertical bends, the resultant unbalanced force is upward or downward and requires the use of a gravity thrust block. Gravity thrust blocks are designed so that the weight of the block offsets the unbalanced upward force in the vertical direction. For downward vertical bends, the thrust block should be an adequate size to ensure that the bearing strength of the soil is not exceeded.

The volume of the gravity thrust block can be calculated by the following formula:

$$\text{Volume} = \frac{T}{B}(SF) = \frac{2PA \sin \frac{\Delta}{2}}{B}(SF) \qquad (4\text{-}33)$$

Where:

T, P, A = as before
SF = safety factor
B = density of block material, lb/ft³
Δ = angle of deflection, degrees

The buoyant density of the thrust block material must be used if it is anticipated that the soil could become saturated or if the water table is above the thrust block elevation.

Example 1: Thrust block design

Determine the thrust block design for a pipe with the following characteristics:

- Pipe = AWWA C900 Pressure Class 150
- Size = 8 in.
- Assumed depth to bottom of thrust block, H_t = 7 ft
- Fitting = 45° horizontal bend
- Soil = sand and gravel
- Design pressure = 150 psi
- Safety factor = 1.5

The unbalanced force required to be overcome by the thrust block is determined from Eq 4-30:

T = $2PA \sin \Delta/2$
D = pipe outside diameter at the joint = 9.050 in.

Therefore, the internal area A is calculated as follows:

$$A = \frac{\pi D^2}{4}$$

$$= \frac{(3.14)(9.050)^2}{4}$$

$$= 64.33 \text{ in.}^2$$

$$T = 2(150)64.33 \sin \frac{45°}{2}$$

$$= 7{,}385 \text{ lb}$$

Thus the unbalanced force is calculated using Eq 4-30 and is shown as follows:

$$R = 2(150)64.33 \sin \frac{45°}{2} = 7{,}385 \text{ lb}$$

From Table 4-11, the soil-bearing capacity is 1,500 lb/ft^2. Therefore, the required design thrust block area can be calculated using Eq 4-31 and is shown as follows:

$$\text{Area} = \frac{T}{S_b}(SF)$$

$$= \frac{7{,}385}{1{,}500}(1.5)$$

$$= 7.39 \text{ ft}^2$$

rounded to 8 ft^2

Use a thrust block with a height H of 2 ft and a length L of 4 ft. Therefore, bearing capacity governs because H is less than 0.5 H_t (i.e., 2 < 0.5 (7)).

Example 2: Thrust block design

For the same situation as Example 1, use a passive resistance thrust block design. Assume H_t is equal to 4 ft.

As with Example 1, the unbalanced force is 7,385 lb.

The required area of the passive resistance thrust block is calculated using Eq 4-32:

$$A = \frac{T(SF)}{\gamma H_t N_d + 2C_s \sqrt{N_d}}$$

Where:

A = required thrust block area, ft^2
T = unbalanced force, 7,385 lb
SF = safety factor, use 1.5
γ = unit weight of soil, 110 lb/ft^3 unsaturated
H_t = height from bottom of thrust block to ground surface (4 ft in this case)
N_d = $\tan^2 (45° + \phi/2)$
ϕ = soil internal friction angle, 30° for sand and gravel
C_s = soil cohesion, 0 lb/ft^2 for sand and gravel

Therefore, calculating the area:

$$A = \frac{7{,}385(1.5)}{110(4) \tan^2\left(45 + \frac{30}{2}\right) + 0}$$

$$= 8.39 \text{ ft}^2$$
round to 9 ft²

$$\tan^2 \alpha = \frac{1 - \cos 2\alpha}{1 + \cos 2\alpha}$$

Use a thrust block with a height H of 2.0 ft and a length L of 4.5 ft.

Restrained Joints—General

When a pipeline is subjected to internal pressures, the resultant thrust forces on the pipe joints can cause joint separation. These thrust forces are a result of the same principles that cause a hydraulic cylinder to operate. Pressure acting on an area results in a force equal to the pressure of the fluid multiplied by the cross-sectional area over which the pressure acts ($F = PA$). In a hydraulic cylinder, the area is the cross-sectional area of the piston based on its outside diameter. In the case of joints in a pipeline where the gasket is carried by the bell, the thrust force is the product of the line pressure and the cross-sectional area based on the pipe outside diameter at the joint.

In straight runs of pipe, the thrust force at each joint is balanced by the reaction at adjacent joints and, in underground systems, by the frictional resistance between the pipe and surrounding soil. With a change in direction of the pipeline, the thrust picture changes dramatically. At a bend, the forces on adjacent joints do not balance each other but combine to create a resultant force that tends to push the bend away from the pipeline. In the case of a horizontal bend, as illustrated in Figure 4-7 with a bend angle Δ, the resultant thrust is given by the equation:

$$T = 2PA \sin (\Delta/2) \tag{4-34}$$

Where:

T = resultant thrust force, lb
P = internal pressure, psi
A = cross-sectional area of the pipe based on outside diameter, in.²
Δ = angle of deflection, degrees

As the angle Δ increases, the resulting thrust also increases. The thrust at a 90° bend is 1.8 times the thrust at a 45° bend and 3.6 times the thrust at a 22½° bend.

When mechanical joint restraint devices are used with fabricated fittings, consult the fitting manufacturer for tensile thrust limitations.

When restrained joints are used, the pipeline becomes its own thrust block. By restraining certain joints at bends and along the pipeline, the resultant thrust force is transferred to the surrounding soil by the pipeline itself. In a properly designed pipeline using restrained joints, the bearing strength of the soil and the frictional resistance between the pipe and soil balance the thrust forces.

Only joint restraint devices manufactured and tested for use in PVC pressure piping systems should be considered. All devices should be required to conform to ASTM F1674, *Standard Test Method for Joint Restraint Products for Use with PVC Pipe*. These joint restraint devices are also suitable for use on PVCO pipe.

When time, material, and labor costs are examined, restrained joints may be more cost effective. There is no requirement to build forms, keep the trench open while the thrust blocks cure, or delay testing of the pipeline. In crowded urban areas, the trench could be opened, the pipe and fitting installed, and the trench closed in the same day. Experience has shown that in a properly designed piping system, concrete thrust blocks can be entirely eliminated.

48 PVC PIPE—DESIGN AND INSTALLATION

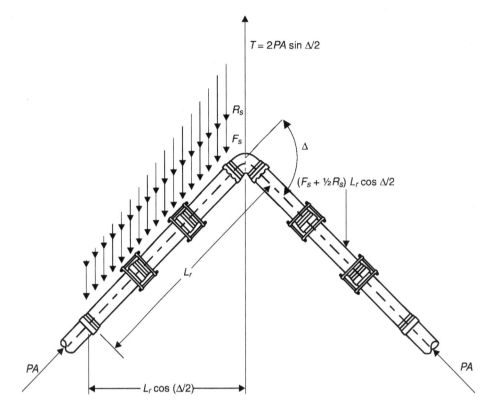

Figure 4-8 Resultant frictional and passive pressure forces on a pipe bend

Resistance to the Unbalanced Thrust Force T

Resistance to the unbalanced thrust force T is generated by the passive resistance of the soil as the bend tries to move, developing resistance in the same manner as a concrete thrust block. In addition to the passive resistance, friction between the pipe and soil also generates resistance to joint separation. In some instances, the passive soil restraint component can be eliminated by other utility work or excavation adjacent to the water pipeline. When this is anticipated, the designer can depend only on the friction between the soil and the pipe for restraint of the pipe and fittings.

Figure 4-8 is a free-body diagram of a restrained horizontal pipeline-bend system designed to resist the unbalanced thrust created by the change in direction. Notice that the thrust T is resisted by the passive resistance R_s as well as the frictional resistance F_s along the length L_r on each side of the bend. L_r is the required length of pipe to be restrained. Note that every joint or fitting must be restrained on both sides of the bend. On small-diameter pipelines, L_r is often less than a full length of pipe. With planning, only the fitting has to be restrained.

As shown in Figure 4-8, the following equation can be used to calculate L_r.

$$L_r = \frac{PA \tan(\Delta/2)}{F_s + 1/2 R_s}(SF) \tag{4-35}$$

Where:

SF = safety factor
F_s = pipe to soil friction, lb/ft
R_s = bearing resistance of the soil along the pipe, lb/ft

Pipe to soil friction F_s. The frictional force acting to oppose movement of the pipeline is a function of the internal shear strength of the soil in relation to the roughness of the pipe's surface.

Internal shear strength of a soil can be expressed by the Coulomb equation:

$$S = C + N \tan(\phi) \tag{4-36}$$

Where:

S = shear strength of the soil, lb/ft^2

C = cohesion of the soil (zero intercept of a plot of shear strength versus normal force), lb/ft^2

N = normal force (force acting perpendicular to the plane of shearing), lb/ft^2

$\tan(\phi)$ = slope of the straight line plot of shear strength versus normal force

ϕ = angle of internal friction of the soil

Potyondy performed a series of investigations to study shearing resistance (skin friction) at the soil/surface interface for various construction materials in contact with different soils. His conclusions define the shearing resistance at the soil/surface interface in terms of a variation on the Coulomb equation using modified values of cohesion and the internal friction angle ϕ. Potyondy redefined these values as constants for a particular material/soil interface within a given moisture range.

Potyondy's equation for the shear strength of a soil/material surface interface is as follows:

$$S_p = f_c C + N \tan(f_\phi \phi) \tag{4-37}$$

Where:

S_p = shearing strength of the surface-to-soil interface, lb/ft^2

f_c = proportionality constant relating the cohesion of a series of direct shear tests on the surface-to-soil interface and the cohesion intercept of the soil

f_ϕ = proportionality constant relating friction angle of a direct shear series (surface/soil) interface to ϕ

Potyondy's study was based on surfaces commonly used for piles (i.e., concrete, wood, smooth and rusted steel). In the early uses of Potyondy's work for the design of pipeline restraint systems, the assumption was made that pipe surfaces were roughly equivalent from a frictional standpoint to rusted steel surfaces, and therefore constants of proportionality recommended by Potyondy for these surfaces were used. However, tests that were made on actual PVC pipe surfaces indicate that values chosen for the original design were not applicable to PVC pipe. It was also found in these studies that the values of f_c and f_ϕ vary with moisture content and the plasticity index of each soil. The selection of f_c and f_ϕ for use in this manual are based on the tests on actual pipe surfaces. Table 4-12 is a tabulation of these values using trench conditions illustrated in Figure 4-9. Applying Potyondy's concepts, the pipe-to-soil friction is defined as

$$F_s = A_p (f_c C) + W \tan(f_\phi \phi), \text{ lb/ft} \tag{4-38}$$

Where:

A_p = area of pipe surface bearing against the soil (horizontal bends: one half the pipe circumference), ft^2

W = normal force per unit length, lb/ft

= $2W_e + W_p + W_w$

$2W_e$ = vertical load on top and bottom surfaces of the pipe taken as the prism load, lb/ft

$W_p + W_w$ = weight of pipe plus weight of water, lb/ft

Table 4-12 Properties of soils used for bedding to calculate F_s and R_s

Soil Group*	f_c	C (lb/ft^2)	f_ϕ	ϕ (deg)	γ (lb/ft^3)	K_n Trench Type		
						3	4	5
GW & SW	0	0	0.7	35	110	.60	.85	1.00
GP & SP	0	0	0.7	31	110	.60	.85	1.00
GM & SM	0	0	0.6	30	110	.60	.85	1.00
GC & SC	0.2	225	0.6	25	100	.60	.85	1.00
CL	0.3	250	0.5	20	100	.60	.85	1.00
ML	0	0	0.5	29	100	.60	.85	1.00

*Soil group per ASTM D2487 (Table 4-6).

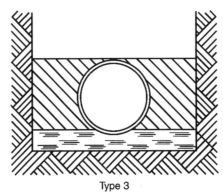

Type 3
Pipe bedded in 4-in. (100-mm) minimum loose soil*
with backfill lightly consolidated to top of pipe.

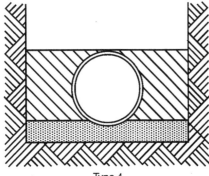

Type 4
Pipe bedded in sand, gravel, or crushed stone
to a depth of ⅛ pipe diameter, 4-in.
(100-mm) minimum, with backfill compacted
to top of pipe. (Approximately 80 percent
Standard Proctor, AASHTO T-99.)

Type 5
Pipe bedded to its centerline in compacted
granular material, 4-in. (100-mm) minimum
under pipe. Compacted granular or select
material* to top of pipe. (Approximately
90 percent Standard Proctor, AASHTO T-99.)

*Loose soil or select material is defined as native soil excavated from the trench, free of rocks, foreign materials, and frozen earth.

Figure 4-9 Suggested trench conditions for restrained joints on PVC pipelines

For design purposes, friction values should always be based on the soil used for bedding. Native soil friction values should be used only when the native soil is also the bedding material.

Bearing Resistance R_s

In addition to friction along the pipe, the resultant thrust is also resisted by the passive pressure of the soil as the pipe tends to move into the surrounding soil. The passive pressure of the soil is generated by the movement. The maximum resistance to this movement can be calculated with the Rankine Passive Pressure formula. The amount of movement required to generate the resistance depends on the compressibility of the soil. In general, soils having a standard proctor density of 80 percent or greater require very little movement to generate the maximum passive resistance of the soil. Because the compressibility of the soil can vary greatly between the suggested trench types shown in Figure 4-9, the design value of passive pressure should be modified by an empirical constant K_n to ensure that excessive movement does not occur. The number of restrained joints required can be minimized by specifying trench type 4 or 5 shown in Figure 4-9.

Rankine's Passive Pressure formula is given by:

$$P_p = (\gamma)H_c N_\phi + C_s Q \qquad (4\text{-}39)$$

Where:

P_p = passive pressure of the soil, lb/ft^2

γ = soil density (backfill density for loose soil, native soil density for compacted bedding), lb/ft^3

H_c = mean depth from surface to plane of resistance (centerline of pipe), ft

ϕ = internal friction angle of the soil

C_s = cohesion of the soil, lb/ft^2

N_ϕ = tan^2 (45° ϕ/2)

Q = tan (45° ϕ/2)

Therefore, the passive resistance on a pipe is defined as

$$R_s = K_n P_p D \qquad (4\text{-}40)$$

Where:

R_s = passive resistance, lb/ft

K_n = trench compaction factor

D = outside diameter of the pipe, ft

Note that soils in the CL and ML groups must be monitored closely because moisture content is difficult to monitor during compaction. Free-draining soils are much better pipe bedding material. Soils in the MH, CH, OL, OH, and PT groups are not recommended for restrained pipe bedding.

Pipelines laid in highly plastic soils subject to high moisture contents are usually bedded in some type of granular material. In cases where the bedding material has a higher bearing value than the native soil, the value of F_s should be calculated using the bedding material values, while the value of R_s should be based on the native soil. In this case, the undrained shear strength values should be used for cohesion according to the "Phi = 0" principle. Actual values of the vane shear tests (AASHTO T223-76), unconfined compression tests (ASTM D2166), or standard penetration test (ASTM D1586) should be used when available.

Table 4-13 In situ values of soil properties for R_s

Soil Group*	$C = S_u$ (lb/ft^2)	γ (lb/ft^3)	K_n Trench Type		
			3	4	5
CL	450	100	.60	.85	1.00
CH	400	100	.40	.60	.85
ML	300	100	.60	.85	1.00
MH	250	100	.40	.60	.85

*Soil group per ASTM D2487.
The above values are for undisturbed soil.

The values in Table 4-13 are for near saturated, undisturbed soils, type CL, ML, CH, and MH, with the pipe surrounded with sand or gravel having a minimum standard proctor density of 80 percent or greater. While these values are conservative for most situations, a competent soils engineer should be contacted for pipelines in wetlands and river bottoms.

AWWA MANUAL M23

Chapter **5**

Pressure Capacity

As a flexible thermoplastic conduit, PVC pressure pipe is primarily designed to withstand internal pressure. PVC pipe and fittings display a unique response to common stress loadings when compared with traditional rigid pipe products. The reliability of the design is based on an assessment of the long-term strength of the PVC compound formulation. Similarly, the design of injection-molded PVC fittings is based on the long-term strength of the fittings themselves. In most instances, the internal pressure controls the PVC pipe design, and the external loads are not a controlling factor. Other influences, such as temperature and the duration and frequency of stress application, may also play a part in the design, but are not first-order parameters. This chapter addresses design relative to internal hydrostatic pressure and dynamic surge pressure.

INTERNAL HYDROSTATIC PRESSURE

PVC pipe, when manufactured for pressure applications, is rated for pressure capacity in accordance with its application and the applicable pressure pipe standards. In North America, PVC pipe is rated for pressure capacity at 73.4°F (23°C). The pressure capacity of PVC pipe is significantly related to its operating temperature (see Table 5-1). As the operating temperature falls below 73.4°F (23°C), the pressure capacity of PVC pipe increases to a level higher than its pressure rating or class. In practice, this increase is treated as an unstated addition to the working safety factor but is not otherwise considered in the design process. Conversely, as the operating temperature rises above 73.4°F (23°C), the pressure capacity of PVC pipe decreases to a level below its pressure rating or class. Table 5-1 shows the response of PVC pressure pipe to change in operating temperature. Anticipated operating temperature is a critical factor that must be considered in the proper design of a PVC pressure piping system. The hydrostatic pressure capacity of PVC pipe is temperature dependent.

The hydrostatic pressure capacity of PVC pipe is also time dependent. The duration of a given hydrostatic pressure application must be considered in the design of a PVC pressure piping system. Although it is the long-term strength of PVC pipe that governs its design, the pipe is inherently capable of withstanding high short-term

Table 5-1 Thermal de-rating factors for PVC pressure pipes and fittings

Maximum Service Temperature		Multiply the Pressure Rating or Pressure Class at 73.4°F (23°C) by These Factors	
°F	(°C)	PVC	PVCO
80	(27)	0.88	.87
90	(32)	0.75	.75
100	(38)	0.62	.64
110	(43)	0.50	.53
120	(49)	0.40	.42
130	(54)	0.30	.31
140	(60)	0.22	—

Notes: The maximum recommended service temperature for PVC pressure pipe and fittings is 140°F (60°C) while that for PVCO pressure pipe is 130°F (54°C).

Interpolate between the temperatures listed to calculate other factors.

Elastomeric compounds used in the manufacture of pipe gaskets are generally suitable for use in water with only slight reduction in long-term properties at continuous temperatures listed above.

The de-rating factors in Table 5-1 assume sustained elevated service temperatures. When the contents of a buried PVC pressure pipe are only intermittently and temporarily raised above the service temperature shown, a further de-rating may not be needed.

pressure surges, regardless of the steady-state pressure. Traditional nonplastic pressure pipes display an insignificant difference between short-term and long-term design strength. A pressure rating for some nonplastic pipes based on quick-burst testing is satisfactory. However, the hydrostatic pressure capacity of PVC pipe, as defined by its pressure rating or pressure class, is derived through long-term hydrostatic pressure testing conducted to establish long-term strength.

Design for long-term, steady-state operating conditions based upon the short-term strength of PVC pipe would be inappropriate. For example, Pressure Class 150 (PC 150) PVC pipe (AWWA C900) will easily withstand a short-term application of 755 psi (5.21 MPa) hydrostatic pressure. However, application of the same pressure for 1 hr could result in pipe burst. But by maintaining PC 150 at a constant pressure of 150 psi (1.04 MPa), a life in excess of 1,000 years could be expected. The pressure class or pressure rating of the product refers to its anticipated steady-state, continuous operating conditions and thus must be derived from its long-term strength. The pipe's ability to withstand short-term applications of substantially higher test pressures will be useful in investigating the pipe's capacity to withstand short-term pressure surges.

The time dependence of PVC pipe response to applied internal hydrostatic pressure can be better understood when consideration is given to the creep property common to all thermoplastic products. PVC pipe, as a thermoplastic product, responds to internal hydrostatic pressure in a manner that is substantially affected by plastic flow or creep. The creep property of PVC pipe is unfortunately misunderstood by many users and engineers. It is not a destructive force that relentlessly undermines the service of a system, nor is creep a property unique to thermoplastic materials. Materials as common as steel (albeit at elevated temperatures) and concrete both experience creep and have to be designed accordingly.

Creep occurs as the PVC pipe responds to an applied stress, such as hydrostatic pressure, with gradual plastic flow or movement. In other words, in response to an applied stress such as internal pressure, the PVC pipe will gradually yield to a point and at a rate that depends on the level and duration of applied stress. The rate of creep in response to a constant internal pressure (or any given stress) decreases with the passage of time. Much data has been accumulated and continues to be accumulated

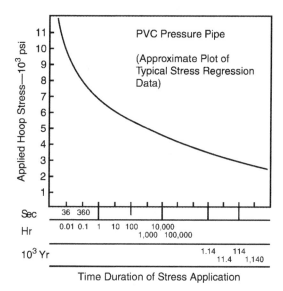

Figure 5-1 Stress regression curve for PVC pressure pipe

proving that PVC pipes held at their pressure ratings continuously for 11.4 years (100,000 hr) will exhibit no perceptible rate of creep at the end of that time. Furthermore, other tests of PVC pipes under constant pressure at twice their pressure ratings (approximately 3 times the pressure class) prove that their quick-burst strength has increased during the 10-year span of the test in spite of the minor creep that has occurred.

Having reviewed the response of PVC pipe to the creep phenomenon, the effect of creep on the performance of PVC pipe in a pressurized water distribution system must be defined. The response of PVC pipe to applied stress is displayed in Figure 5-1 and is known as the stress regression curve (SR curve) for PVC pressure pipe. It is the curve that results when a series of stress/time data (failure) points are joined together. It does not represent a loss of strength in a single pipe with time. The curve, as shown in Figure 5-1, has been plotted with the horizontal axis representing the logarithm of time to permit plotting a greater passage of time—from 10 seconds to 1.1 million years. If the time axis were plotted on a linear scale (Cartesian coordinates), the variation in hoop stress, which relates directly to applied internal hydrostatic internal pressure, would appear insignificant from 100,000 hours to 500 years. In a practical sense, a long-term response to applied hydrostatic pressure can be based on the hoop stress rating at 100,000 hr in that

- The response of the PVC pipe to applied internal hydrostatic pressure or applied hoop stress has essentially stabilized at 100,000 hr when considering the design life of piping systems at 50 to 100 years.

- The response of PVC pipe to applied hoop stress after 100,000 hr can be accurately determined through testing performed in accordance with ASTM D1598, *Standard Method of Test for Time-to-Failure of Plastic Pipe Under Long-Term Hydrostatic Pressure*, and through analysis performed as required in ASTM D2837, *Standard Method for Obtaining Hydrostatic Design Basis for Thermoplastic Materials*.

The pressure class or pressure rating for all PVC pipe manufactured in North America is based on a refined plot of stress regression, commonly termed the stress regression line (SR line). Figure 5-2 is a typical SR line for PVC pressure pipe. The SR

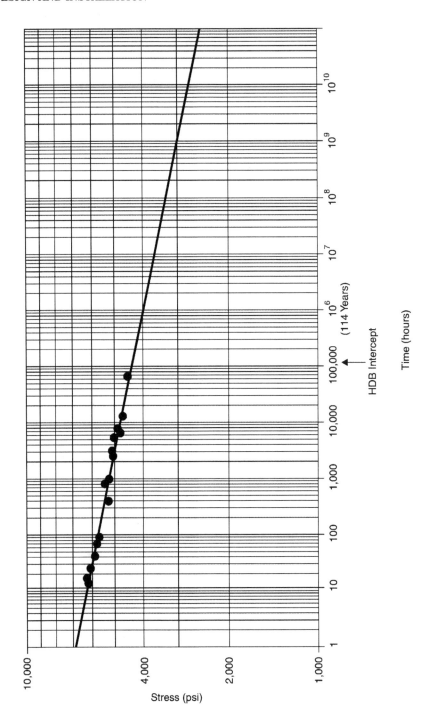

Figure 5-2 Stress regression line

line is a plot of the same data plotted on the SR curve; however, the hoop stress on the vertical axis is also plotted with logarithmic scale when preparing the SR line. The log-log plot of long-term stress response data for PVC pipe plots as a straight line. Use of the SR line, with constant slope, permits accurate mathematical projection of extremely long-term response. It is possible to project that the estimated life for a PVC pressure pipe, operating at full-rated pressure (i.e., 2,000 psi [13.79 MPa] hoop stress) is between 1×10^{16} or 1×10^{17} hr, or between 1.14×10^{12} and 1.14×10^{13} years.

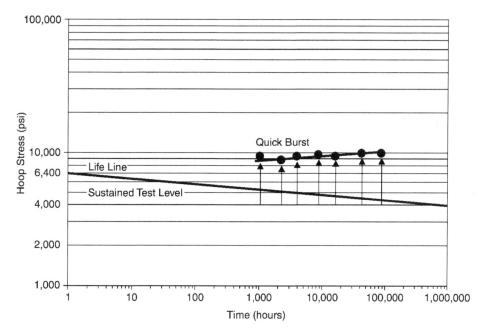

Figure 5-3   Strength and life lines of PVC 12454

Research published in 1981 by Robert T. Hucks Jr. helped confirm this long-term strength behavior while dramatically demonstrating the retention of strength throughout the life period. Figure 5-3 is a log-log plot of the quick-burst hoop stresses as they were recorded by Hucks. All pipe specimens were held at a constant hoop stress of 4,000 psi (27.58 MPa). Actual quick-burst values increased from about 8,000 psi (55.16 MPa) to over 10,000 psi (68.95 MPa).

All PVC pressure pipe manufactured in North America must be extruded from PVC compounds for which stress regression lines have been established. PVC pipe designed and manufactured for pressure water applications must have a defined hydrostatic design basis (*HDB*) of 4,000 psi (27.58 MPa) as defined by ASTM D2837. PVCO pipe designed and manufactured for pressure water applications must have a defined *HDB* of 7,100 psi (49.0 MPa) as defined by ASTM D2837. Examples using 4,000 psi (27.58 MPa) will be used throughout this manual. However, it should be recognized that other *HDB*s may be used for specific products such as injection-molded PVC pressure fittings.

When submitting a PVC pressure pipe extrusion or injection-molding compound for approval, a manufacturer must establish the SR line and *HDB* for the PVC material through long-term hydrostatic pressure testing in accordance with ASTM D1598 and ASTM D2837. Having established that a PVC pressure pipe extrusion compound provides an *HDB* equal to or greater than 4,000 psi (27.58 MPa) or 7,100 psi (49.0 MPa) for PVCO, the compound can then be assigned a hydrostatic design stress and is qualified in long-term stress rating for the manufacture of PVC pressure pipe.

When qualifying a joint design, a manufacturer has two options. The first option is to thicken the bell to maintain the *DR* of the pipe. The second option is to test to verify that joint assemblies qualify for an *HDB* category of 4,000 psi (27.58 MPa).

In the definition of PVC pipe's pressure rating, the hydrostatic design stress, *S*, is used in the calculations, rather than the *HDB*. The design stress is simply the value obtained when *HDB* is divided by a safety factor *SF*.

$$S = \frac{HDB}{SF} \tag{5-1}$$

58   PVC PIPE—DESIGN AND INSTALLATION

Where:

S = design stress, psi
HDB = hydrostatic design basis, psi
SF = safety factor

The design stress S serves as the maximum hoop stress value used in calculation of PVC pipe pressure rating. PVC pipe pressure rating is calculated in accordance with standard practice defined by the International Standards Organization (ISO) Equation R161-1960 based on the work first published by Lamé in 1852:

$$\frac{2S}{P} = \frac{D_o}{t} - 1 = SDR - 1 \qquad (5\text{-}2)$$

Where:

S = design stress, psi
P = pressure rating, psi
D_o = average outside diameter, in.
t = minimum wall thickness, in.
SDR = D_o/t, standard dimension ratio (also called DR)

The ISO Equation can be transposed to a form whereby it can be better utilized in defining pressure capacities of PVC pipe:

$$P = \frac{2S}{DR - 1} \qquad (5\text{-}3)$$

The above equation can be used at any time for a given DR to determine what internal pressure would create a given hoop stress. Or conversely, the equation can determine the magnitude of hoop stress generated by a given internal pressure.

The specific design procedure for PVC pressure pipe is dependent upon its particular watermain application—either distribution or transmission. Distribution is defined as piping whose main purpose is to act essentially as a header in distributing water through numerous and frequent lateral connections. An example would be piping serving a typical residential development with service connections every 100 ft. Transmission is defined as piping whose main function is to supply water directly from one location to another. An example would be a supply line from a water source to a reservoir from where the water would perhaps enter the distribution system.

The design approach for each is presented below.

DISTRIBUTION MAINS

The following is a procedure for designing PVC pressure pipe within the looped perimeter of a distribution system. Operating conditions are typically low velocity (about 2 ft/sec) with frequent but minor pressure surges due to numerous service connections. AWWA C900 addresses this specific application for pipe sized from 4 in. through 12 in., Cast-Iron Outside Diameter.

C900 PVC pipe is manufactured in three pressure classes. These are PC 100, PC 150, and PC 200. The pressure class formula is as follows:

$$PC = \frac{2S}{DR - 1} - P_s \qquad (5\text{-}4)$$

Where:

PC = pressure class, psi
S = design stress, psi

PRESSURE CAPACITY   59

Table 5-2   Pressure classes of PVC pipe (C900)

DR	Pressure Class, PC (*psi*)
25	100
18	150
14	200

DR = dimension ratio

P_s = surge for 2 ft/sec velocity, psi

The design stress for this distribution application is derived using the *HDB* of PVC 12454 and a factor of safety of 2.5. Specifically, the design stress will be

$$S = HDB/SF$$

$$= 4,000 \text{ psi}/2.5 = 1,600 \text{ psi}$$

The term P_s and its derivation is discussed in greater detail later is this chapter. The values for P_s used in the PC formula can be found by multiplying the values of P_s' presented in Table 5-6 by two.

In other words,

$$P_s = 2 \times P_s' \tag{5-5}$$

Where:

P_s' = surge for 1 ft/sec velocity change

Example: Determine pressure class (PC) of PVC *DR* 18 for distribution applications.

Solution: Use Eqs 5-4 and 5-5.

Given that $S = 1,600$ psi, $P_s' = 17.4$ psi (from Table 5-6)

$$PC = \frac{(2)(1,600 \text{ psi})}{18-1} - (2)(17.4 \text{ psi})$$

$$= 153 \text{ psi} \cong 150 \text{ psi}$$

Thus, *DR* 18 PVC is assigned a pressure class of 150 psi for a distribution application.

A summary of the three pressure classes presently available from AWWA C900 is given in Table 5-2.

For PVCO pipe, pressure classes are derived in exactly the same fashion. The only differences lie in the material property values. PVCO has an *HDB* = 7,100 psi and the values for surge allowance, P_s, are 31, 27, and 23 psi, respectively, for Pressure Classes 200, 150, and 100. In addition, the term D_o/t is substituted for *DR* in Eq. 5-4.

TRANSMISSION MAINS

The application of PVC pressure pipe for water transmission mains differs greatly from that of a distribution main. There are virtually no cyclic stresses, velocities may be higher, and surges in general are very occasional.

Pipe conforming to AWWA C905 in sizes 14 in. through 48 in. (350 mm through 1,200 mm) is most often used in this application. Even if AWWA C900 pipe is used as a transmission main (rather than looped inside a distribution system), the design criteria presented below should be used.

First, the design stress, S, must again be determined using the formula below:

$$S = HDB/F$$

Table 5-3 Pressure ratings of PVC pipe (C905)

DR	PR (psi)
51	80
41	100
32.5	125
26	160
25	165
21	200
18	235
14	305

For AWWA pipe, PVC with a minimum cell classification of 12454 must be used with the HDB = 4,000 psi. In AWWA C905, the safety factor F' = 2.0.

Thus, for a transmission main application,

$$S = 4{,}000 \text{ psi}/2.0 = 2{,}000 \text{ psi}$$

AWWA C905 uses the terminology *pressure rating* to identify long-term, steady-state pressure capabilities of different DRs of PVC. The pressure rating (PR) is calculated using the following formula:

$$PR = 2S/(DR - 1) \tag{5-6}$$

Using the value of 2,000 psi for S, and by substituting the different DRs available in AWWA C905, the pressure ratings are listed in Table 5-3.

These PRs are one of the two upper limits for steady-state operating pressures (WPR being the other) within each DR of a PVC transmission main.

To design for occasional pressure surges, the following equation is used:

$$WPR = STR - (V \times P_s') \tag{5-7}$$

Where:

WPR = working pressure rating, psi
STR = short-term rating of pipe, psi
V = actual system velocity (ft/sec)
P_s' = one ft/sec surge pressure, psi—from Table 5-6

The STR is calculated by applying a factor of safety (SF') to the short-term strength of PVC pipe, which is defined to be 6,400 psi hoop stress. PVC pipe including the gasketed joint must be able to withstand this stress as part of conformance to ASTM D3139. The ISO Equation is then used to convert the maximum hoop stress to short-term pressure strengths (STS) for each DR available as presented in Table 5-4.

By applying a safety factor F' = 2.5, and by rounding, the following STRs are obtained using the formula from Eq 5-8. The results are presented in Table 5-5.

$$STR = STS/SF' \tag{5-8}$$

These levels should be considered to be the design surge capacity limits for PVC pressure pipe manufactured to AWWA standards for a transmission main application. By limiting positive pressure surges to these levels, a minimum safety factor of 2.5 will be maintained against the short-term strength of the PVC material. The values of STR also represent a level of approximately 25 percent above the pressure rating for each DR of pipe.

Table 5-4 Short-term strengths of PVC pipe

DR	STS (psi)
51	256
41	320
32.5	406
26	512
25	533
21	640
18	753
14	985

Table 5-5 Short-term ratings of PVC pipe

DR	STR (psi)
51	100
41	130
32.5	165
26	205
25	215
21	255
18	300
14	395

After a WPR has been computed for a particular DR, it should be compared to the PR of that same DR. The lower of the two values should be used as the upper limit for the steady-state operating pressure of the system. In that way, the designer can maintain a minimum safety factor of 2.0 between the operating pressure and the long-term hydrostatic strength (HDB), as well as a minimum safety factor of 2.5 between the worst-case total pressure (i.e., operating plus surge) and the STS of the pipe.

The actual safety factor against surge (F') can be calculated as follows:

$$SF' = STS/P_{max} \qquad (5\text{-}9)$$

Where:

P_{max} = maximum possible total pressure inside pipe

= operating pressure (P_{op}) + ($V \times P_s'$)

Example: For PVC transmission pipe operating at 140 psi and at a velocity of 3.5 fps, select the proper DR of PVC pipe and compute the safety factor against surge.

Solution:

$$DR = (2S/P) + 1$$
$$= (2)(2{,}000 \text{ psi})/140 \text{ psi} + 1$$
$$= 29.6 \text{ (maximum)}$$

Thus, try DR 25

(a) PR = 165 psi, STR = 215 psi

compute WPR

$$WPR = 215 \text{ psi} - (3.5 \times 14.7 \text{ psi})$$
$$= 164 \text{ psi}$$

Therefore, because the operating pressure (140 psi) does not exceed the *PR* (165 psi) nor the *WPR* (164 psi) of this system, *DR* 25 is selected.

(b) compute safety factor for surges, F'

First calculate P_{max},

$$P_{max} = P_{op} + V \times P_s'$$
$$= 140 \text{ psi} + (3.5 \times 14.7 \text{ psi})$$
$$= 191.5 \text{ psi}$$
$$SF' = STS/P_{max}$$
$$= 533 \text{ psi}/191.5 \text{ psi}$$
$$= 2.78$$

INJECTION-MOLDED PVC FITTINGS

The design of gasketed, injection-molded PVC pressure fittings meeting AWWA C907 and designed for use with Class 100 or Class 150 AWWA C900 PVC pressure pipe uses the same principles stated in the AWWA pipe standards. The central idea is that the pressure capacity of PVC pressure pipe and fittings must be based on long-term pressure tests.

The long-term pressure tests required for PVC pipe design as previously described in this chapter result in a stress regression curve illustrated in Figure 5-1. The data from which the stress regression curve is determined are derived from tests performed on extruded small-diameter pipe specimens made from the PVC compound being examined. The resulting *HDB* can be applied to the design of any PVC pressure pipe made with that compound.

For injection-molded PVC fittings, the long-term performance is less specifically dependent on the PVC compound used in their manufacture. Within the limits allowed by the standard, a manufacturer can choose to make the fittings from one of a number of compounds and employ a variety of machines and mold designs. The total effect of these manufacturing decisions is specific to one size and type of fitting. Because of this, the long-term hydrostatic pressure tests are performed on the fittings themselves rather than the PVC compound. Each size of each configuration of fitting is subjected to stress regression analysis in the same manner as pipe compounds, except that internal pressure is substituted for hoop stress.

The requirement is that the fitting must have proven long-term pressure strength equivalent to that of Pressure Class 150 PVC pipe. The pressure strength of Pressure Class 150 PVC pipe can be calculated from a modified form of Eq 5-3.

$$SF \times LTPS = \frac{2 \times HDB}{DR - 1} \tag{5-10}$$

Where:

$LTPS$ = long-term pressure strength of pipe projected to 100,000 hr, i.e., the pressure equivalent of the *HDB* of the pipe compound

SF = safety factor, which for this calculation = 1

DR = 18 for Pressure Class 150 pipe

HDB = hydrostatic design basis, 4,000 psi for PVC pressure pipe

Performing this calculation yields a long-term pressure strength of 470 psi.

The normal pressure class of injection-molded fittings meeting AWWA C907 can be calculated from Eq 5-11, which is very similar to the one used to calculate the pressure class of pipe.

The relationship is

$$PC = \frac{LTPS}{SF} - P_s \qquad (5\text{-}11)$$

Where:

PC = pressure class, psi
$LTPS$ = 470 psi
SF = 2.5
P_s = pressure rise in DR 18 Pressure Class 150 PVC pipe as a result of an instantaneous stoppage of a 2 ft/sec flow = 35 psi

Therefore,

$$PC = \frac{470}{2.5} - 35 = 150$$

The nominal pressure class of the fittings therefore equals 150 psi at 73.4°F (23°C).

Because the fittings are designed for use with cast-iron outside diameter pipes, their usual application will be in Pressure Class 150 or Pressure Class 100 PVC pipe systems (i.e., pipes conforming to AWWA C900). The fittings may also be used in systems of pressure rated (Series) PVC pipes having iron pipe size-outside diameters by using a special transition gasket in the bell. Because the fittings were designed around usage of DR 18 PVC pipe, the fittings may be used with IPS Series pipe having DRs equal to or greater than DR 18. Specifically, the fittings are suitable to be used with DR 21, DR 26, DR 32.5, or DR 41 Series PVC pipe.

If the PVC fittings will be operated at temperatures above 73.4°F (23°C), the pressure class should be reduced accordingly by applying the appropriate factor from Table 5-1.

FABRICATED PVC FITTINGS

PVC pressure fittings for use with PVC pressure pipe may also be fabricated out of sections of PVC pipe. The design of these fittings for internal pressure is identical to that for pipe. The appropriate pressure class or pressure rating of pipe should first be determined depending on the application and operating conditions. The pressure class or rating selected will dictate the particular DR of PVC pipe. If fabricated fittings are to be included in the design, they should be made from pipe having the same DR as the pipe required to obtain the desired pressure class or rating. For example, if a designer determines that PR 165 (DR 25) pipe is suitable for a transmission main, any fabricated fittings should be manufactured from DR 25 PVC pipe. Designers may also choose to select one DR thicker than that of the pipe for the added structural safety factor where extreme earth loads are a design concern. The pipe should also be designed for large earth loads.

More specific details on the manufacturing and testing processes for fabricated PVC fittings are contained in AWWA C900 and C905.

DYNAMIC SURGE PRESSURE

In a general sense, surge pressures are any deviation from the normal steady-state hydrostatic pressure in a piping system. Normally, positive surges are considered; however, negative surges to the vapor pressure of the pipe contents may occur. Both should be brought to acceptable levels by the use of suitable protection devices or operating procedures. There are certain key concepts that should be familiar to those who design, install, test, and operate piping systems.

Surge pressures or *water hammer* are generated in any pressurized piping system if the flowing liquid changes velocity. When flow velocity changes, part of all of the kinetic energy of the moving fluid must be converted to potential (stored) energy and ultimately dissipated through frictional losses in the fluid or in the pipe wall if the fluid is to regain its original pressure. Some of the more common causes of pressure surges are as follows:

- The opening and closing (full or partial) of valves
- Improperly sized pressure-reducing valves
- Starting and stopping of pumps
- Changes in turbine speeds
- Changes in reservoir elevation
- Reservoir wave action
- Liquid column separation
- Entrapped air

Surges may generally be divided into two categories: transient surges and cyclic surges. Transients may best be described as the intermediate conditions that exist in a system as it moves from one steady-state condition to another. The closing of a single valve is a typical example. Cyclic surging is a condition that recurs regularly in time. Surging of this type is often associated with the action of equipment such as reciprocating pumps and pressure-reducing valves. Small oscillatory surges can grow rapidly in magnitude and can become damaged if the frequency is at or near the natural resonant frequency (harmonic) of the piping system. Any piping material including PVC may eventually fatigue if exposed to continuous cyclic surging at sufficiently high frequency and stress amplitude.

In a distribution network consisting of AWWA C900 or C909 pipes and fittings made from either iron, molded PVC, or fabricated PVC, specific surge calculations are not easily accomplished. However, the built-in surge allowances are adequate for operating conditions where the flow is maintained at or below 2 fps. In water transmission pipelines where a relatively simpler layout exists, the designer can investigate transient potential of the specific system operating conditions using the *WPR* analysis.

Transient Surges

The magnitude of individual or transient surges can be calculated reliably using the elastic wave theory of surge analysis. The pipeline designer should be aware that the geometry and boundary conditions of many systems are complicated and require the use of refined techniques similar to those developed by Streeter and Wylie.

Analysis of transient surge pressures is demonstrated in the calculation of the pressure rise in a pipeline as a result of the rapid closing of a valve. The pipeline is assumed to be supported against longitudinal movement and is equipped with expansion joints. The maximum surge pressure is related to the maximum rate of flow, while the rate of travel of the pressure wave is related to the speed of sound in the fluid (modified by the piping material).

The wave velocity is given by the following equation:

$$a = \frac{4{,}660}{\sqrt{1 + (kD_i / Et)}} \tag{5-12}$$

Where:

a = wave velocity, ft/sec

k = fluid bulk modulus (300,000 psi for water)

D_i = pipe inside diameter, in.

E = Modulus of elasticity of the pipe material (400,000 psi for PVC, 465,000 psi for PVCO)

t = wall thickness, in.

Substituting dimension ratio, DR for $(D_i/t + 2)$
[Note that $DR = D_o/t = (D_i + 2t)/t = D_i/t + 2$]

$$a = \frac{4{,}660}{\sqrt{1 + (k(DR-2)/E)}} \quad (5\text{-}13)$$

The maximum pressure surge may be calculated using

$$P = \frac{aV}{(2.31)g} \quad (5\text{-}14)$$

Where:

V = maximum velocity change, ft/sec^2

g = acceleration due to gravity, 32.2 ft/sec^2

P = pressure surge, psi

Example: A flow of 2 ft/sec is suddenly stopped in a 6-in. Pressure Class 150 (DR 18) PVC pipe.

Calculate the expected maximum surge pressure.

Solution: First calculate the wave velocity:

Substitute the given values into Eq 5-13

k = 300,000 psi

DR = 18

E = 400,000 psi

$$a = \frac{4{,}660}{\sqrt{1 + (300{,}000(18-2)/400{,}000)}}$$

Solving yields, a = 1,292 ft/sec

Now substitute a and V into Eq 5-14

Solving for P yields $P = \frac{(1{,}292)(2)}{(2.31)(32.2)} = 35$ psi

Pressure surges in PVC pipe of different dimension ratios in response to a 1 ft/sec (0.3 m/sec) instantaneous flow velocity change are shown in Table 5-6.

Larger-diameter (i.e., over 12-in. sizes) transmission pipelines are characterized by fewer lateral connections than are found in distribution networks, generally higher flow velocities and fewer (and thus easily identifiable) barrier effects. It may be advantageous to the designer in subjecting a proposed transmission pipeline to computer software analysis, now commercially available. The response of steady-state, startup, and shutdown operations (both programmed and power failure) can be simulated and displayed graphically or numerically. As well, the duration and location of significant positive and negative pressure surges can be identified.

Using this information, the designer can

- choose the type and proper location for protection devices
- estimate the economics of certain options such as variable speed pumps
- examine the advantages of relocating the pipeline to minimize any positive or negative surge pressures
- make the most cost-effective choice of PVC pipes for each section of the pipeline exhibiting acceptable operating requirements

Table 5-6 PVC pressure surge versus DR for 1 ft/sec (0.3 m/sec) instantaneous flow velocity change

Dimension Ratio, DR	Pressure Surge, P_s'	
	psi	(kPa)
14	19.8	(139)
18	17.4	(120)
21	16.0	(110)
25	14.7	(101)
26	14.4	(99)
32.5	12.8	(88)
41	11.4	(79)
51	10.8	(74)

Cyclic Surges

When transient pressures of noticeable magnitude occur on a frequent basis, cyclic stresses may be imposed upon the pipe material. If the cyclic stresses occur frequently enough at a sufficiently high magnitude, fatigue failure is a possibility in a thermoplastic pipeline, such as PVC.

Cyclic stresses are found more commonly in sewage force mains or turf irrigation lines than in water mains. Smaller-sized distribution mains do generate cyclic surges; however, the generously large safety factors in the pressure class design terminology are thought to eliminate any fatigue failure concerns. For transmission piping, cyclic surges, if they occur at all, typically would be at such infrequent intervals that cyclic design is not warranted in the vast majority of cases.

Should the designer wish to investigate the effects of cyclic surges, reference should be made to the *Handbook of PVC Pipe* published by the Uni-Bell PVC Pipe Association.

The design process should involve consideration of each parameter discussed in this chapter. Because the nature of transmission pipelines permits interactive analysis, the designer can minimize costs without sacrificing sound engineering judgment.

Common Surge Pressure Control Techniques

As a result of the wide variety of surge conditions possible, positive or negative pressures, transient or oscillatory, there is no general solution applicable to the control of surge conditions for every pipeline. However, the following techniques have been found to be useful in a multitude of cases.

Surge tank or air chamber. A closed unit containing air and water, sometimes separated by a diaphragm or bladder. The air is under pressure allowing control of both positive and negative surges in high-pressure systems by allowing flow into and out of the unit.

Standpipe. A tank open to the atmosphere that functions in a manner similar to a surge tank for low pressure.

Surge tank with one-way outlet. A surge tank which allows water to enter the line during negative surges and allows no return on positive surges (useful for negative pressures only).

Pump and motor flywheels. Pumps and motors with flywheels decelerate slowly and minimize surges generated.

PRESSURE CAPACITY 67

Slow-closing valves. These are adjustable and may be mechanically or hydraulically operated to minimize abrupt variation in flow velocity.

Pressure-relief or by-pass valves. Spring loaded or hydraulically operated valves which release and vent flow in excess of a preset pressure value.

Time delays. Wired into the pump control circuit, these can prevent excessive cycling or pump startup during a high-amplitude oscillation.

TRANSMISSION PIPE DESIGN EXAMPLE

This analysis of a relatively simple pipeline will illustrate the use of the design principles discussed in this chapter. PVC pipe standards offer a variety of pipe strengths and sizes. Ideally, the designer will make selections that minimize capital and operating costs while maintaining an adequate design safety factor.

The project is a 20,000 ft long PVC water transmission main designed for an ultimate capacity of 4,000 gpm (5.76 mgd).

The profile of the pipeline is shown in Figure 5-4. Water is being pumped to a ground storage tank (point f) with a maximum water level of 35 ft from the floor. The centerline of the discharge end of the main, at the tie-in to the storage tank, will be 5 ft below the tank floor.

Key stations and their elevations along the pipeline are

Point	Station	Elevation at Pipe Centerline (ft)
a	0 + 00	600
b	45 + 00	670
c	75 + 00	720
d	115 + 00	800
e	165 + 00	940
f	200 + 00	940

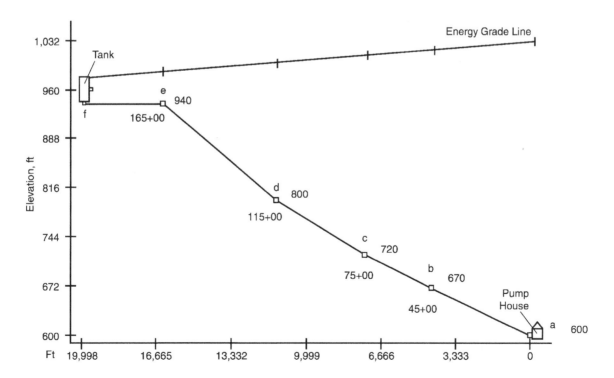

Figure 5-4 Pipeline profile

The objective of the design process will be to select proper DRs of PVC for appropriate sections of pipeline while never exceeding the PR nor the WPR of the pipe at any point. An effort will be made to select DRs that meet the design criteria while providing optimum economic value for the utility or owner.

The key determinant of PVC pressure pipe design is the internal pressure. The pipe dimensions can be found in the AWWA pipe standards. For this example, AWWA Standard C905, *Polyvinyl Chloride (PVC) Pressure Pipe and Fabricated Fittings, 14 In. Through 48 In. (350 mm Through 1,200 mm), for Water Transmission and Distribution*, was used. The exact pipe dimensions are required to determine the flow velocity. The total pressure in the pipeline at any point is the sum of the static head, the friction loss, and the pressure rise as a result of sudden velocity changes. For simplicity, the selection of PVC pipe in this example will be limited to four PRs in CIOD only (PR 235, 165, 125, and 100).

Step 1: Determine the maximum flow velocity.

Assume that 20-in. PVC pipe will be used. In AWWA C905, the heaviest wall shown to be available in 20-in. pipe is DR 18. The assumption of beginning with the heaviest wall (i.e., the lowest DR) is recommended for most designs in the initial stage. The first assumption may be confirmed or revised as the design is developed.

$$\text{Average ID} = \text{Average OD} - 2 \,(\text{minimum wall thickness} \times 1.06)$$

NOTE: The tolerance on wall thickness is +12 percent approximately. There is no minus tolerance. Manufacturers will generally aim to produce PVC pressure pipe with wall thicknesses about 6 percent over the minimum.

Assume:
20-in. DR 18 per AWWA C905

$$\text{Avg. ID} = 21.60 - 2\,(1.200 \times 1.06)$$
$$= 19.05 \text{ in.} = 1.59 \text{ ft}$$
$$V = Q/A$$

Where:
Q = flow in ft^3/sec = 4,000 gpm or 8.91 ft^3/sec
A = area, ft^2
V = velocity, ft/sec

$$A = (3.14)\,(1.59/2)^2 = 1.98 \text{ ft}^2$$

Therefore, $V = 8.91/1.98 = 4.5$ ft/sec

Because the velocity is within an acceptable range, the design may proceed with 20-in. pipe.

Step 2: Determine the surge factor.

In a transmission pipeline, the amplitude and location of the surge pressure envelope will often be analyzed by computer. For this example, the assumption has been made that the maximum surge pressure will be equal to an instantaneous stoppage of flow at full velocity. In practice, the costs of pipe materials may be significantly reduced through the use of appropriate surge control devices and proper pipeline operating procedures.

The pressure rise resulting from a $V = 4.5$ ft/sec instantaneous velocity change in PVC pressure pipes can be charted as follows:

Dimension Ratio, DR	1 ft/sec Surge, P_s' (psi)	$V \times P_s'$ (psi)
41	11.4	51.3
32.5	12.8	57.6
25	14.7	66.2
18	17.4	78.3

Step 3: Determine the WPR for each of the DRs of Step 2.

The WPR is a job-specific pressure rating of the pipe, taking into account the maximum possible surges versus the short-term strength of the pipe. The WPR may be either higher or lower than the PR of the pipe, depending on the flow conditions. The lower value between the WPR and the PR should be used as the upper limit for system steady-state operating pressure.

$$WPR = STR - V \times P_s'$$

DR	STR (psi)	$V \times P_s'$ (psi)	WPR (psi)	PR (psi)
41	130	51.3	78.7	100
32.5	165	57.6	107.4	125
25	215	66.2	148.8	165
18	300	78.3	221.7	235

It can be seen that the governing parameter for the pressure design of this example will be the WPR analysis since it is lower than the PR of each DR.

Step 4: Determine the friction loss f under full-flow conditions.

Continue to assume DR 18 for this calculation because this pipe will produce slightly greater losses than the other DRs under consideration. The result will be conservative for all design operations.

Information on friction losses is given in chapter 3, Hydraulics. The Hazen–Williams equation is convenient to use:

$$f = 0.2083\,(100/C)^{1.852}\,\frac{Q^{1.852}}{d_i^{4.8655}}$$

Where:

f = friction head, ft of water per 100 ft of pipe

d_i = inside diameter of pipe, in.

Q = flow, gpm

C = flow coefficient, 150 for PVC

Substituting for 20-in. PR 235 pipe, where d = 19.05 in.

f = 0.273 ft of water per 100 ft of pipeline

= 0.118 psi per 100 ft (station) of pipeline

Step 5: Determine the pressures at key points in the pipeline under steady-state, full-flow conditions.

This pressure, P, at any point is the sum of the static head as a result of difference in elevations and the friction loss.

Referring to Figure 5-4, the pressure at key points can be calculated as follows:

Starting at the storage tank:

Sta. 200 + 00			
Static Head	= 980 –940	=	40 ft
or 40 ft × (0.43 psi/ft)		=	17.3 psi
Sta. 165 + 00			
Static Head	= (980 –940) ft × (0.43 psi/ft)	=	17.3 psi
Friction Head	= (3,500 ft) × (0.118 psi/100 ft)	=	<u>4.1</u> psi
Total Head		=	21.4 psi

Sta. 115 + 00
    Static Head   = (980 –800) ft × (0.43 psi/ft)   =   77.4 psi
    Friction Head   = (8,500 ft) × (0.118 psi/100 ft)   =   <u>10.0</u> psi
    Total Head   =   87.4 psi

Sta. 75 + 00
    Static Head   = (980 –720) ft × (0.43 psi/ft)   =   111.8 psi
    Friction Head   = (12,500 ft) × (0.118 psi/100 ft)   =   <u>14.8</u> psi
    Total Head   =   126.6 psi

Sta. 45 + 00
    Static Head   = (980 –670) ft × (0.43 psi/ft)   =   133.3 psi
    Friction Head   = (15,500 ft) × (0.118 psi/100 ft)   =   <u>18.3</u> psi
    Total Head   =   151.6 psi

Sta. 0 + 00
    Static Head   = (980 –600) ft × (0.43 psi/ft)   =   163.4 psi
    Friction Head   = (20,000 ft) × (0.118 psi/100 ft)   =   <u>23.6</u> psi
    Total Head   =   187.0 psi

The pressure, P, at each of the key points are summarized as follows:

Point	Station	Static Head (psi)	Friction Head (psi)	Pressure, P (psi)
f	200 + 00	17.3	0	17.3
e	165 + 00	17.3	4.1	21.4
d	115 + 00	77.4	10.0	87.4
c	75 + 00	111.8	14.8	126.6
b	45 + 00	133.3	18.3	151.6
a	0 + 00	163.4	23.6	187.0

Step 6: Determine the appropriate DR of pipe for each section of the pipeline.

From previous calculations in Step 3, DR 18 PVC pressure pipe has a working pressure rating 221.7 psi. For the next greater DR, DR 25, the WPR is 148.8 psi. Therefore, DR 18 is selected to start out at the pumphouse until a point in the system where the operating pressure, P, drops to be equal to the WPR of DR 25. At this point, DR 25 may be used. Subsequent steps will determine the starting points for DR 32.5 as well as DR 41.

It can be seen from the above summary of pressures by section that the transition to DR 25 will occur between Stations 45 + 00 and 75 + 00, in section bc. To pinpoint the exact location, the pressure gradient for that section must be calculated.

$$\Delta P(bc) = \frac{Pc - Pb}{\text{Sta. Length of bc}}$$

$$= \frac{126.6 \text{ psi} - 151.6 \text{ psi}}{(75-45) \times (100 \text{ ft})}$$

$$= -0.83 \text{ psi}/100 \text{ ft}$$

The length beyond Sta. 45 + 00 (point b) can be calculated as follows:

$$\text{Sta. Length} = \frac{WPR(DR\ 25) - Pb}{\Delta P(bc)}$$

$$= \frac{(148.8 \text{ psi}) - (151.6 \text{ psi})}{-0.83 \text{ psi}/\ 100 \text{ ft}}$$

$$= 337 \text{ ft (i.e., at 337 ft downstream of Sta. 45 + 00)}$$

PRESSURE CAPACITY   71

Therefore, begin using *DR* 25 at Sta. 48 + 37.

Similarly, the transition point for *DR* 32.5 can be found.

From the summary of pressures and knowing the *WPR* of *DR* 32.5 is 107.4 psi, *DR* 32.5 can be used between Sta. 75 + 00 and Sta. 115 + 00, i.e., section cd.

First, calculate the pressure gradient in section cd.

$$\Delta P(cd) = \frac{P_d - P_c}{\text{Sta. Length of cd}}$$

$$= \frac{87.4 \text{ psi} - 126.6 \text{ psi}}{(115 - 75) \times (100 \text{ ft})}$$

$$= -0.98 \text{ psi}/100 \text{ ft}$$

Next, the Station Length beyond Sta. 75 + 00 can be calculated:

$$\text{Sta. Length} = \frac{WPR(DR\ 32.5) - P_c}{\Delta P(cd)}$$

$$= \frac{(107.4 \text{ psi}) - (126.6 \text{ psi})}{-0.98 \text{ psi}/100 \text{ ft}}$$

$$= 1{,}959 \text{ ft (i.e., at 1,959 ft downstream of Sta. 75 + 00)}$$

Therefore, may begin using *DR* 32.5 at Sta. 94 + 59.

Similarly, it can be calculated where *DR* 41 usage may begin.

From review of the summary of pressures and knowing that the *WPR* of *DR* 41 for this example is 78.7 psi, the range for *DR* 41 begins between Sta. 115 + 00 and Sta. 165 + 00, i.e., section de.

First, calculate the pressure gradient in section de.

$$\Delta P(de) = \frac{P_e - P_d}{\text{Sta. Length of de}}$$

$$= \frac{21.4 \text{ psi} - 87.4 \text{ psi}}{(165 - 115) \times (100 \text{ ft})}$$

$$= -1.32 \text{ psi}/100 \text{ ft}$$

Next, the Station Length beyond point *d*:

$$\text{Sta. Length} = \frac{WPR(DR\ 41) - P_d}{\Delta P(de)}$$

$$= \frac{(78.7 \text{ psi}) - (87.4 \text{ psi})}{-1.32 \text{ psi}/100 \text{ ft}}$$

$$= 659 \text{ ft (i.e., at 659 ft downstream of Sta. 115 + 00)}$$

Therefore, *DR* 41 may begin usage at Sta. 121 + 59 and continue for the duration of the pipeline up to its terminus at the reservoir.

The design for internal pressure may be summarized as follows:

Distance from Pumphouse (*ft*)	Use 20 in.	Pressure Gradient (*psi*)
0–4,837	*DR* 18 (*PR* 235)	187.0–148.8
4,837–9,459	*DR* 25 (*PR* 165)	148.8–107.4
9,459–12,159	*DR* 32.5 (*PR* 125)	107.4–78.7
12,159–20,000	*DR* 41 (*PR* 100)	78.7–17.3

In this example of a 3.8 mile pipeline, the designer has the opportunity to make significant cost savings through the use of several PVC pipe pressure ratings. Computer modeling may disclose even further potential savings by showing exactly where and how surge control is most effective. (Note that the above pipe selection was made assuming that the potential exists for the instantaneous stoppage of flow.)

If the pipeline is operated in a cycle mode (i.e., like some sewage force mains), an analysis of fatigue life should be made. Both present and future modes of operation should be examined.

Although the effect of external loads (earth and traffic) is seldom a design consideration for pressure pipes, certain situations (e.g., deep burial and high DR) may control the pipe design. For details of superimposed loads and flexible pipe design, see chapter 4.

AWWA MANUAL M23

Chapter 6

Receiving, Storage, and Handling

This chapter details procedures for receiving PVC and PVCO pipe, including inspection, unloading, and recommended practices for subsequent storage and handling.

RECEIVING

When receiving the PVC pipe shipment at the job site, the contractor or purchaser should adhere to the following procedures, which are suggested as common practices.

Inspection

Each pipe shipment should be inventoried and inspected upon arrival, even though the pipe should have been inspected and loaded at the factory using methods acceptable to the carrier. The carrier has the responsibility to deliver the shipment in good condition. The receiver has the responsibility to ensure that there has been no loss or damage. The records that accompany each shipment should provide a complete list of all items shipped.

The following procedures for acceptance of delivery are recommended:

- Examine the load. If the load is intact, inspection while unloading should be sufficient to ensure that the pipe has arrived in good condition.

- If the load has shifted, has broken packaging, or shows rough treatment, then each piece should be carefully inspected for damage.

- If small-diameter pipe is "nested" in large-diameter pipe for shipments, it should be inspected carefully to ensure that damage caused by shifting pipe has not occurred.

- Check total quantities of each item against shipping records (pipe gaskets, fittings, lubricant, etc.).

- Note any damaged or missing items on the delivery receipt. This includes any damage caused by severe diesel sooting on the tractor side of shipment. NOTE: covering load with tarp will prevent this.

- Notify carrier immediately of any damage or loss and file a claim in accordance with their instructions.

- Do not dispose of any damaged material. The carrier will notify you of the procedure to follow.

- Replacements for shortages and damaged materials are normally not shipped without request. If replacement material is needed, reorder from the manufacturer, the distributor, or their representative.

Unloading

The means by which PVC pipe is unloaded in the field is the decision and responsibility of the receiver. Preferred unloading is in package units using mechanical equipment; however, the pipe can be unloaded individually by hand.

When unloading package units, the following instructions should be carefully followed:

- Remove restraints from the top unit loads. These may be either fabric or steel straps, ropes, or chains with padded protection.

- If there are boards across the top and down the sides of the load, which are not part of the pipe packaging, remove them.

- Use a forklift (or front-end loader equipped with forks) to remove each top unit from the truck, one at a time. Remove units on the rear end of the truck bed first. Do not run the forks too far under units, as fork ends striking adjacent units may cause damage.

- If a forklift is not available, then a spreader bar may be used with fabric straps that are capable of handling the load. The straps should be spaced approximately 8 ft (2.4 m) apart and looped under the load. Cables also may be used if cushioned with rubber-hose sleeves or other material to prevent abrasion of the pipe.

- To unload lower units, repeat the above unloading process.

CAUTION:

- During removal and handling, be sure that the units do not strike anything.

- Severe impact could cause damage, particularly during cold weather.

- Do not handle units with individual chains or single cables, even if padded.

- Do not attach cables to unit frames or banding for lifting.

- Pipe package units should be stored and placed on level ground. Package units should not be stacked more than 8-ft high and should be protected by dunnage in the same way that they were protected while loaded on the truck.

If unloading equipment is not available, pipe may be unloaded by removing individual pieces by hand. However, care should be taken to ensure that pipe is not dropped or damaged (see Figure 6-1).

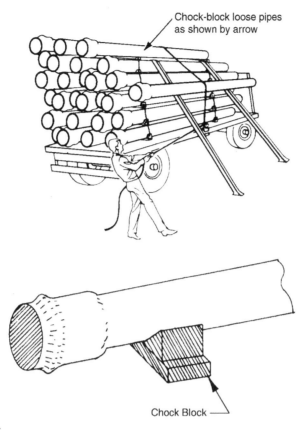

Figure 6-1   Chock block

STORAGE

A problem commonly experienced on pipe construction projects is the damage to piping products during storage. The following procedures and practices are recommended to prevent damage to PVC pipe during storage:

- If possible, pipe should be stored at the job site in unit packages provided by the manufacturer. Caution should be exercised to avoid compression, damage, or deformation to bell ends of the pipe. NOTE: Normally, PVC pipe in packages will display bell ends arranged alternately with pipe spigots.

- When unit packages of PVC pipe are stacked, the weight of the upper units should not cause deformation to pipe in lower units.

- PVC pipe unit packages should be supported by racks or dunnage to prevent damage to the bottom during storage. Supports should be spaced to prevent pipe bending.

- When long-term storage (more than 2 years) with exposure to direct sunlight is unavoidable, PVC pipe should be covered with an opaque material. Adequate air circulation above and around the pipe should be provided as required to prevent excessive heat accumulation (see Environmental Effects—Thermal Effects in chapter 1).

- PVC pipe should not be stored close to heat sources or hot objects, such as heaters, boilers, steam lines, or engine exhaust.

- When unit packages of PVC pipe are stacked, ensure that the height of the stack does not result in instability, which could cause stack collapse with resultant pipe damage or personal injury.
- The interior and all sealing surfaces of pipe, fittings, and other appurtenances should be kept free of dirt and foreign matter.
- Gaskets should be protected from excessive exposure to heat, direct sunlight, oil, grease, and ozone.

AWWA MANUAL M23

Chapter **7**

Installation

SCOPE

This chapter presents and discusses the practices and requirements for the installation of PVC water distribution, PVCO water distribution, and PVC transmission water mains. The information presented in this chapter is not a substitute for site-specific installation conditions or contractual obligations. The installation procedure should be selected based on a wide range of considerations, many of which are discussed in this chapter.

It is important to understand that installation is one of the most important aspects of a project. A good design can be ruined as a result of poor installation.

ALIGNMENT AND GRADE

For most projects, the principal reference points are provided by the engineer or owner. These principal reference points must be based on a permanent datum. The horizontal layout should be tied to a coordinate system in addition to property lines. Vertical control should be based on geodetic elevations. Traverses should be closed for horizontal control, and benchmarks should be checked for vertical control.

The transfer of the line and grade of the excavation work from control points established by the engineer is usually the responsibility of the contractor. This is a critical part of the project and must be checked thoroughly.

All pipes should be laid to and maintained at the established lines and grades. Fittings, valves, air-release valves, and hydrants should be installed at the required locations with the valve and hydrant stems plumb.

Prior to starting work on the project, the entire site should be subjected to a thorough reconnaissance survey. The purpose of this survey is to identify potential problems during construction, which may include size and operating space requirements of equipment, intensity of traffic, and location of trees and overhead wires.

INSTALLATION IN TRENCHES

Most water mains are installed in a trench, which is a relatively narrow excavation dug in undisturbed soil. However, it is very easy for a trench to become an embankment. It is

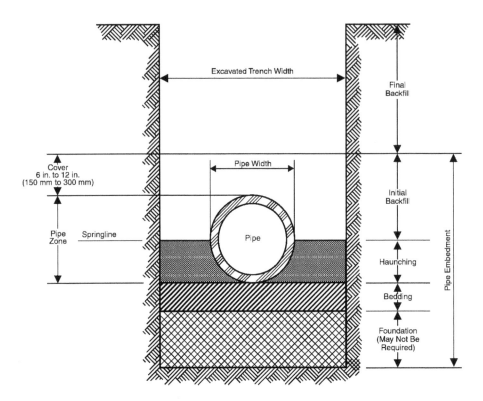

Figure 7-1 Trench cross section showing terminology

important that the language used in the design, in the contract, and at the construction site be consistent and correct. The terminology commonly used in PVC installation practice is shown on Figure 7-1 and defined in the following text.

Terminology

Foundation. A trench foundation lies beneath the bedding and is required when the native trench bottom is unstable.

Bedding. The bedding is directly underneath the pipe and brings the trench bottom to grade. The purpose of the bedding is to provide a firm, stable, and uniform support of the pipe.

Haunching. The haunching area begins at the bottom of the pipe and ends at the springline of the pipe. This area is the most important in terms of limiting pipe deflection.

Initial backfill. Initial backfill begins above the springline of the pipe to a level of 6 in.–12 in. (150 mm–200 mm) above the top of the pipe.

Final backfill. Final backfill begins above the initial backfill to a level below that required for the trench area restoration.

After the pipe is placed in the trench, the trench is backfilled to the original ground surface. The load on the pipe will develop as the backfill settles because the density of the backfill is less than that of the native material. The resultant earth load on the pipe is equal to the weight of the material above the pipe, less the shearing or friction forces that will reduce the settlement between the trench material and the original soils. (See chapter 4.)

Trench Width

As the width of the trench increases, the effect of contributing support of the backfill earth load on the pipe is reduced. At a width defined as the *transition width*, this

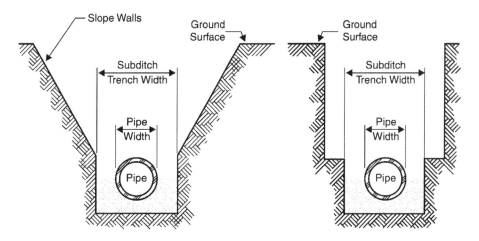

Figure 7-2 Examples of subditches

shearing force has no effect on the earth load on the pipe. An excavation with a width greater than the transition width is considered an embankment in terms of pipe design.

The pipe designer and site representative must discuss the design aspects before beginning project construction. The implication of trench width on the design of the pipe must be considered. If the earth load design used the Marston's theories, as outlined in chapter 4, then the design may be sensitive to trench width. However, if the soil prism theory was used, then the design is not affected by the trench width.

The trench width at the ground surface may vary with and depend on its depth, the nature of soils encountered, and the proximity of adjacent structures. If compaction of pipe bedding and haunching are deemed necessary, the minimum trench width as measured at the springline of the pipe should be 18 in. (450 mm) or the pipe outside diameter plus 12 in. (300 mm), whichever is greater. The trench width must always provide sufficient room between the trench wall and the pipe to ensure that compaction will occur.

An alternative to the narrow, vertical-walled trench is to lay the pipe in a subditch and backcut or slope the sides of the excavation above the top of the pipe. Two examples of this are shown in Figure 7-2.

Wide trenches are more properly called embankments. In wide trenches depending on the loading, it is important that the pipe embedment be compacted at least two-pipe diameters beyond the outside of each side of the pipe. The entire trench or embankment must be properly backfilled and compacted if settlement at the surface needs to be prevented.

It is necessary to provide a trench protection system where required by statute, in locations where poor soils are present during construction, in congested areas that have many utilities which may require support during construction, or in confined working areas. Under certain circumstances, the trench protection system may increase the load on the pipe to a greater amount than the design load.

If sheeting is used as a trench protection system and left in place, the coefficient of side friction may be reduced. As discussed in previous chapters, the load coefficient C_d would increase, therefore increasing the load on the pipe.

If the sheeting is to be removed, then it should be removed as the trench backfilling takes place. This will allow the frictional forces to develop between the sides of the trench and the backfill materials. If the sheeting is removed after the backfill has been completed, then much of the backfill material will retain its shape and will not fill the void left by removing the sheeting. This void will eliminate the frictional contact between the backfill and the trench wall and result in an increased load on the pipe.

The contractor should be responsible for designing, installing, and removing any trench support systems. The design of the trench support system should be provided by a professional engineer. The geotechnical report may provide some parameters that may be used for the design of the trench support systems. Many jurisdictions have safety requirements that must be followed, and the people involved in the construction supervision and administration should be familiar with these requirements.

Excavating

The excavation for a trench must be dug to the approximate grade of the pipe with a width as shown in the construction drawings. The difficulty in excavating is greatly influenced by the depth, native soil materials, presence of rock, water table, rural or urban locations, other utilities, and traffic requirements.

The depth of the excavation required to install the pipe can be determined from the design drawings. The contractor must have adequate equipment onsite to handle the anticipated pipe depth. Some additional excavation may be required under the pipe because of a poor foundation, and the excavating equipment used should have sufficient capacity to handle poor foundation conditions.

The native soil materials and the presence of rock at the site can be anticipated from the geotechnical report prepared for the project. This report should be kept onsite at all times for reference. The actual soils encountered should be regularly recorded in the site diary to document the project and to assist in the review of possible claims as a result of changed conditions.

Dewatering

Dewatering or predrainage of construction sites is usually carried out by vacuum wellpoints, deep wells, or eductor wells. The function of dewatering is to efficiently remove only enough water from an aquifer to facilitate underground construction. The choice of dewatering method depends on aquifer characteristics, depth of required drawdown, site geometry, and cost.

Vacuum wellpoint system. A vacuum wellpoint system relies on atmospheric pressure to remove the groundwater upward to the ground surface where it can be collected or discharged. Typically, vacuum wellpoint systems can achieve 20 ft (6.6 m) of drawdown at the wellpoint line. As the distance increases from the wellpoint, the amount of drawdown will depend on the aquifer characteristics.

The wellpoint system includes a vacuum unit, which is used to create low air pressure in the header-wellpoint systems. As a result of the low pressure at the wellpoint, atmospheric pressure forces water up to the surface. The groundwater is then collected at the surface in a separation tank and discharged to an appropriate location.

Deep-well system. Deep wells are used to dewater the ground where there is a large wetted depth below the excavation level and pervious soils exist, resulting in high well yields. Deep wells are the simplest and oldest method of removing water from the ground. In a deep-well system, the groundwater flows by gravity to a well where it is removed by pumping. The pumping unit is usually a submersible turbine pump. If a large amount of water is to be pumped, vertical line shaft pumps with the motors at ground surface are used.

Eductor system. The eductor system is usually used for dewatering fine-grained soil where the depth of dewatering is greater than 20 ft (6.6 m).

An eductor system can employ either a single or double pipe design. Single pipe systems consist of a large-diameter pipe that forms the well casing with a small inner pipe that forms the return line. Water is pumped under high pressure through the annular space between the two pipes and is forced through a nozzle and venturi. This causes low pressure to develop around the nozzle. The low pressure will draw water to the well and up the return line. Typical operating pressures are 7–10 atmospheres on the pressure side.

Eductor systems typically have a limited quantity of water that can be pumped, usually ranging from 3 to 11 gpm (0.19 to 0.69 L/s). An advantage of the eductor system is that the vacuum is developed directly at the well screen; with a properly installed seal above the sand pack around the wellpoint, a considerable vacuum can be transmitted to the soil.

Dewatering systems in general. The ideal dewatering system will lower the groundwater to just below the level in a trench. The depth of the dewatering system and its capacity are designed so that the water level is at sufficient depth below the excavation at the opposite side of the trench. In some cases, a single-line dewatering system may not lower the water table to a level below the trench on the opposite side of the excavation. In these cases, dewatering systems must be installed on both sides of the trench.

Where stratified conditions occur, groundwater may enter the trench from a confined pervious aquifer when the dewatering system is installed on only one side of the trench. A sand wick is also needed to drain the confined layers above the well screen.

Wellpoint spacing is a function of the depth of required drawdown, soil permeability, and the presence of impermeable layers.

The general contractor and his dewatering subcontractor are usually assigned the responsibility for deciding upon the need, design, operation, maintenance, and removal of the dewatering systems. The owner is responsible for providing adequate information on the ground conditions. The information should include detailed borehole logs, where groundwater was encountered during drilling, water level in piezometers sealed into specific water bearing stratum, and size distribution curves of the water-bearing soils.

Owners should recognize the importance of the dewatering aspects of a project. Claims for extra work or damages to adjacent structures often arise from insufficient geotechnical information and particularly from a lack of appreciation of the severity of the groundwater problem by both the owner and the contractor.

Stockpiling Excavated Materials

Excavated material should be stockpiled in a manner that will not endanger the work activities or obstruct sidewalks, driveways, or public safety devices. The stockpiled material must not damage trees and lawns.

Stored material should be kept back from the wall of the trench to prevent loading, which could lead to a cave-in.

Fire hydrants under pressure, valve pit covers, valve boxes, curb stop boxes, fire and police call boxes, or other utility controls should remain unobstructed and accessible. Gutters and ditches shall remain clear unless other arrangements have been made for drainage. Natural watercourses should not be obstructed and an attempt shall be made to try to anticipate what might happen during a rainstorm or extended exposure to the elements.

Preparation of Trench Foundation

The trench foundation should be constructed to provide a firm, stable, and uniform support for the full length of the pipe. Trench excavation below the pipe bottom followed by the placement of granular bedding material is both economical, practical, and usually reliable.

Where a soft trench foundation is encountered, overexcavation and backfilling with granular material may be necessary for stabilization. The depth of the stabilization material should be determined by tests and observations and a geotechnical consultant may be required.

Caution must be used when constructing a stabilized trench foundation. Uniformly graded crushed stone should not be used in wet conditions without being encased in a geotextile fabric that can prevent the entry of the subgrade particles into the foundation or bedding material. If this intrusion occurs, excessive pipe settlement may take place. In cases where a trench foundation cannot be stabilized with bedding material and where intermittent areas of unequal settlement are anticipated, special foundations for the pipe may be necessary.

All bedrock, boulders, cobbles, and large stones should be removed from the excavation to provide at least 4 in. (100 mm) of embedment cushion on each side of and below the pipe and appurtenances. The trench foundation should be cleaned of all loose or projecting rocks prior to the placement of the bedding material. These loose pieces of rock, cobbles, and stones may cause point loading on the pipe that could lead to pipe failure.

Trenches in rock sometimes have water flowing along the bottom of the trench. This may wash away the fine materials in the trench which may lead to trench settlement. Clay trench plugs can be used to reduce the flow of water in the trench.

If blasting is required to remove rock, such work must be undertaken by specialists. Blasting is an effective and controlled engineering procedure, but it is also a spectacular and sometimes startling operation which can lead to the formation of false impressions in the mind of an observer, particularly if that observer is also a property owner.

Before blasting is carried out on any site, a careful examination, preferably including photographs of all buildings and structures liable to be within the range of perceptible vibration from the source, should be carried out. If this pre-blast survey is carefully made, all existing cracks and defective conditions can be recorded and their number and locations can be brought to the attention of the owner. Should a claim for damage subsequently arise, a post-operation check can establish which defects have occurred as a result of the blasting operation.

Laying of Pipe

Pipe and accessories should be inspected for defects and cleanliness before they are lowered into the trench. Any defective or damaged material should be repaired or replaced, and all foreign matter or dirt should be removed from the interior of the pipe and accessories before lowering into the trench. When pipe laying is not in progress, open ends of installed pipe should be closed to prevent entrance of trench water, dirt, foreign matter, or small animals into the line.

All pipe, fittings, valves, hydrants, and accessories should be carefully lowered to prevent damage. Pipe and accessories should never be dropped or dumped into the trench.

PIPE JOINTS

Pipe Joint Assembly

The assembly of PVC pipe requires the careful adherence to the proper joint assembly procedures outlined in the contract documents. The following is a suggested procedure:

1. Confirm that the bell and gasket are free from any foreign material that could interfere with the proper assembly of the pipe joint. Some gaskets are

restrained in the bell and should not be removed. Contact the manufacturer for specific information.

2. Confirm that the pipe spigot end is clean. Wipe with a clean, dry cloth around the entire circumference from the pipe end to about 1 in. (25 mm) beyond the reference mark.

3. Lubricate the spigot end of the pipe using a lubricant and method of application recommended by the pipe manufacturer. The entire circumference should be lubricated, especially the bevelled end of the spigot.

4. Do not lubricate the gasket or the gasket groove in the bell because the lubrication could cause gasket displacement. After the spigot end has been lubricated, it must be kept clean and free of dirt and sand. If dirt or sand adhere to the lubricated end, the spigot must be wiped clean and relubricated.

5. If the system is for potable water, the lubricant must be approved for potable water service. Do not use a nonapproved lubricant that may harbor bacteria or damage the gaskets or pipe.

6. Insert the spigot end into the bell so that it is in uniform contact with the gasket. Push the spigot until the reference mark on the spigot end is flush with the end of the bell. The recommended method for assembly is using a bar and block; however, pullers such as a lever or come-along may also be used.

7. If undue resistance to insertion of the spigot end is encountered or the reference mark does not reach the flush position, disassemble the joint and check the position of the rubber gasket. If it is twisted or pushed out of its seat, clean the gasket, bell, and spigot end and repeat the assembly steps. Be sure both pipe lengths are in proper alignment. If the gasket is not out of position, measure the distance between the reference mark and the spigot end and check it against the correct values provided by the pipe manufacturer.

8. When pipe laying is not in progress, the open ends of the installed pipe should be closed to prevent the entrance of trench water into the line. Whenever water is prevented from entering the pipe, enough backfill should be placed on the pipe to prevent floating. Any pipe that has floated shall be removed from the trench and the bedding restored.

Pipe Joint Offset

The gasketed joint design of PVC pipe allows for the minor discrepancies of not placing pipes to proper grade and alignment. There is, however, no minimum offset requirement in the standards. Known changes in alignment and grade should be accomplished with high-deflection couplers or fittings.

As a practical matter, PVC joints may be offset slightly until the point where the outside diameter of the spigot contacts the inside diameter of the bell. Pipe manufacturers should be consulted for the amount of offset allowed. (See Joint Offset in chapter 4.)

PIPE CUTTING AND BENDING

Pipe Cutting

To set a fitting in the desired location, it may be necessary to cut the pipe. A square cut is essential to ensure proper assembly. AWWA C900 and C905 pipe can be easily cut with a fine tooth hacksaw, handsaw, or a portable power saw with a steel blade or abrasive discs. Prior to cutting the pipe, the pipe should be marked around its entire circumference to assist in making a square cut.

Use a factory-finished pipe end as a guide to determine the angle and length of bevel. After cutting and bevelling, a reference mark must be made on the spigot end to ensure that the joint is properly assembled. The reference mark may be located by using a factory-marked end of the same size pipe as a guide. The reference point may also be located by referring to data provided by the pipe or fitting manufacturer. Installers should be aware that when assembling two different brands of PVC pipe, the reference line position may differ. In this case, the correct reference line for the spigot brand "X" is obtained by measuring the insertion depth of an uncut length of Brand "X." For iron fittings (push-on or mechanical joints), remove all but ¼ in. (6 mm) of the factory-made bevel from the spigot end. The PVC pipe spigots should be bottomed in the iron fittings.

Pipe Bending

PVC pipe does offer the advantage of longitudinal bending to accomplish changes of direction. If the pipe is to be bent, precautions must be taken to ensure straight alignment in the joint. Also, the pipe must meet minimum bending radius requirements at every point (see Performance Limits in Longitudinal Bending in chapter 4).

PIPE EMBEDMENT

Embedment Material and Type

The materials available for use as pipe embedment material should be identified, as much as practicable, before construction begins. A change in native soil conditions may require a change in the construction procedure. Figure 7-3 provides typical embedment types listed in AWWA C605, *Underground Installation of Polyvinyl Chloride (PVC) Pressure Pipe and Fittings for Water.*

Installation

PVC pipe should be installed with proper bedding to provide uniform longitudinal support under the pipe. The pipe barrel should be in contact with the trench bottom for its full length to ensure uniform bearing of the pipe. Bell holes may be required. The stiffness of the bedding should match that of the native soil material. If the bedding materials are too stiff, the pipe may be subjected to a bottom line load. The gradation of the pipe bedding material should closely relate to that of the native material to prevent migration into or out of the pipe bedding materials. A suggested minimum thickness of bedding is 4 in. (100 mm).

Careful trimming of the excavated trench bottom to fit the pipe barrel is not essential and very difficult to accomplish with the desired degree of accuracy. Therefore, pipe can be placed on the flat bottom of the trench.

In rock trenches, the minimum thickness of the bedding material should be 4 in. (100 mm) to cushion the pipe and to reduce the possibility of point loading.

Satisfactory installation and compaction of the pipe haunching is an important component of PVC pipe installations. Good compaction of the pipe haunches will result in acceptable pipe deflection. The pipe haunch material must be worked in and compacted to provide complete contact with the pipe bottom and to ensure that there are no voids in the material.

Initial backfill is also important to the successful installation of PVC pipe. During the installation and compaction of the initial backfill, it is important not to damage the pipe.

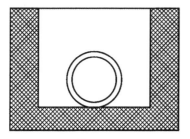

Type 1
Flat-bottom trench.* Loose embedment.
$E' = 50$ psi (340 kPa), $K = 0.110$

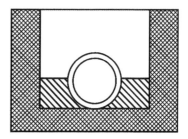

Type 2
Flat-bottom trench.* Embedment lightly
consolidated to centerline of pipe.
$E' = 200$ psi (1,380 kPa), $K = 0.110$

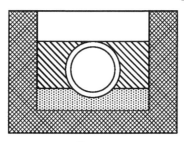

Type 3
Pipe bedded on 4 in. (100 mm) minimum of loose soil.†
Embedment lightly consolidated to top of pipe.
$E' = 400$ psi (2,760 kPa), $K = 0.102$

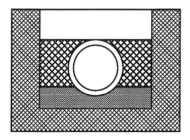

Type 4
Pipe bedded on sand, gravel, or crushed stone
to a depth of ⅛ pipe diameter, 4 in. (100 mm)
minimum. Embedment compacted to top of
pipe. (Approximately 80 percent standard
proctor, AASHTO T-99 or ASTM D698.)
$E' = 1,000$ psi (6,900 kPa), $K = 0.096$

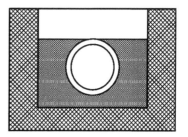

Type 5
Pipe embedded in compacted granular
material to centerline of pipe. Compacted
granular or select material† to top of pipe.
(Approximately 90 percent standard proctor,
AASHTO T-99 or ASTM D698.)
$E' = 2,000$ psi (13,800 kPa), $K = 0.083$

Flat-bottom is defined as undisturbed earth.

†*Loose soil* or *select material* is defined as native soil excavated from the trench, free of rocks, foreign materials, and frozen earth. A soft, "loose soil" bedding will contour to the pipe bottom. Caution must be exercised to ensure proper placement of embedment material under the haunches of the pipe.

Note: Required embedment type will depend on the pipe's dimension ratio, internal operating pressure, and external load, and shall be specified by the purchaser.

Figure 7-3 Recommendations for installation and use of soils and aggregates for foundation, embedment, and backfill

86 PVC PIPE—DESIGN AND INSTALLATION

Hand-held or walk-behind compaction equipment can be used to install the pipe embedment material without damaging the pipe. Larger compaction equipment can be used when the total depth of soil on top of the pipe is equal to or greater than 2 ft (0.6 m).

All pipe embedment material should be selected and placed carefully, avoiding rounded rocks over 1½ in. (40 mm), angular rocks over ¾ in. (20 mm), frozen lumps, and debris.

After placement and compaction of pipe embedment material, the balance of the backfill material may be machine placed. The final backfill materials should not contain large stones or rocks, frozen material, or debris. Proper compaction procedures should be followed to provide the required soil densities specified by the design engineer.

CASINGS

Casings may be required where PVC pipe is installed under highways, runways, or railways for the following reasons:

- To prevent damage to structures caused by soil erosion or settlement in case of pipe failure or leakage
- To permit economical pipe removal and placement in the future
- To accommodate regulations or requirements imposed by public or private owners
- To permit boring rather than excavation where open excavation would be impossible or prohibitively expensive

When PVC pipe is installed in casings, skids may be required to prevent damage to the pipe and bell during installation and to provide proper long-term support. PVC pipe in casings should not rest on bells. Skids should properly position the PVC pipe in the casing. Figure 7-4 shows a typical skid arrangement on PVC pipe.

Skids may extend for the full length of the pipe, with the exception of the bell and spigot position required for assembly, or may be spaced at intervals. Skids must provide sufficient height to permit clearance between bell joint and casing wall. Skids should be fastened securely to pipe with strapping, cables, or clamps. Commercially fabricated casing spacers can also be used (see Figure 7-5).

Table 7-1 provides recommendations on casing size required for different sizes of PVC pipe. Casings are normally sized to provide an inside clearance that is at least 2 in. (50 mm) greater than the maximum outside diameter of the pipe bell, pipe skids, or cradle runners.

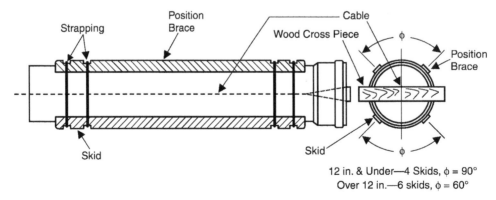

Figure 7-4 PVC pipe casing skids

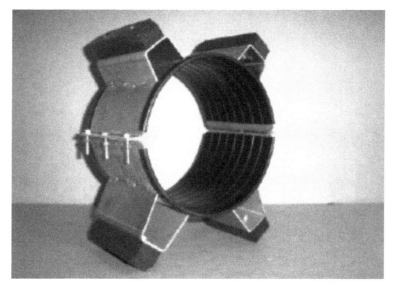

Source: Cascade Waterworks stainless steel casing spacer, style "CCS."

Figure 7-5 Casing spacer

Table 7-1 Recommended casing size

Nominal Pipe Size (Diameter, *in.*)	Casing Size (Inside Diameter)	
	in.	*(mm)*
4	8–10	203–254
6	10–12	254–305
8	14–16	356–406
10	16–18	406–457
12	18–20	457–508
14	24–26	610–660
15	24–26	610–660
16	28–30	711–762
18	30–32	762–813
20	32–34	813–864
21	32–34	813–864
24	36–38	914–965
27	38–40	965–1,016
30	44–48	1,118–1,219
33	46–48	1,168–1,219
36	48–50	1,219–1,270
42	54–56	1,350–1,400
48	60–62	1,500–1,550

Pipe may be installed in the casing using a winch-drawn cable or jacking. In both methods, avoid damage to pipe or bell joints. Use of lubricant (flax soap or drilling mud) between skids and casing can ease installation.

CAUTION: Do not use petroleum products, for example, oil or grease. Prolonged exposure to these products can cause damage to some gaskets. Do not use creosote-treated wood, which can weaken PVC pipe through chemical attack.

Table 7-2 Maximum recommended grouting pressures

DR	Maximum Grouting Pressure* psi	(kPa)
51	6	(41.4)
41	12	(82.7)
32.5	25	(172.4)
26	51	(351.6)
25	58	(399.9)
21	100	(689.5)
18	164	(1,130.7)
17	196	(1,351.4)
14	367	(2,530.4)

* These maximum pressures are based on the temperature in the wall of the pipe not exceeding 73.4° F (23° C). Maximum grouting pressures must be reduced with increased wall temperatures.

Upon completion of pipe insertion, backfilling should be done in accordance with design requirements. During backfilling, floating the PVC pipe out of the proper position should be prevented. When pressure grouting, grouting pressures should be controlled and monitored to prevent distorting or collapsing the pipe. Table 7-2 lists the maximum recommended grouting pressures as a function of the dimension ratio (DR).

APPURTENANCES

Piping systems include pipe and various appurtenances required in the control, operation, and maintenance of the systems. Proper design, installation, and operation of PVC piping systems must relate to appurtenances, as well as pipe.

System Requirements

Control valves. Control valves (gate, ball, or butterfly) must be provided in the water system to permit isolation of any one line within the system. Secondary lines are valved from main feeder lines. In high-value commercial and industrial areas, control valves are normally located at intervals no greater than 500 ft (152 m). In other areas, control valve intervals are normally 800 to 1,200 ft (244 to 366 m).

Safety valves. Pressure relief valves are important in long pipelines for surge control. Air relief valves are desirable at high points in pressure lines where other relief is not available. Blow-off valves are used at low system elevations and dead ends to permit line emptying or flushing when necessary. Vacuum relief valves are used to prevent complications caused by negative pressures.

Fire hydrants. Fire hydrants are normally spaced to provide maximum fire protection coverage of 40,000–160,000 ft^2 (3,700–14,900 m^2), depending on the needed fire flow. Distribution lines servicing fire hydrants are normally 6 in. (150 mm) nominal diameter or larger. Hydrant connections from main lines should be valved.

Fittings. Fittings are required for changes in line direction or size and for branch connections (for example, tee and cross fittings) and are available in a variety of designs and materials. Cast-iron or ductile-iron fittings are sometimes used with cast-iron dimensioned PVC water main.

PVC Fittings

PVC fittings for use in pressure systems can be either injection-molded or fabricated from extruded PVC pressure pipe or machined from extruded PVC pressure pipe. Molded fittings should be in conformance with AWWA C907. Fabricated fittings should be in conformance with AWWA C900 or C905 and should have the same pressure rating (PR) or pressure class (PC) of the pipe used in the system. Fittings of greater PR or PC may also be used.

Currently, there are no fabricated or injection-molded PVCO fittings. For gasketed couplings machined from extruded PVC pipe, the computed dimension ratio shall not be greater than the DR of the pipe used in the system. Exceptions are the annular gasket space and the coupling entry, where the wall thickness shall be at least the minimum wall thickness of the pipe used in the system.

Handling

PVC fittings should be handled with the same degree of care and caution as is recommended for the pipe. Specifically, fittings should never be thrown or tossed, nor subjected to severe impacts which may damage the fitting. Larger fittings may require mechanical assistance for lifting during unloading or lowering into the trench. Canvas straps are recommended as lifting slings.

Assembly

Assembly procedure for gasketed joint PVC fittings are identical to those of regular PVC pipe. The spigot end of pipe should be cut squarely, chamfered, and then lubricated before insertion into the bell. Mechanical assistance may be necessary for joint assembly in the larger sizes. Excavator buckets should never be used to push directly on PVC fittings. They may only be used with canvas slings to pull a fitting into an assembly. Come-along pullers are also widely used and accepted. Specific assembly recommendations should also be obtained from the manufacturer.

Backfill

Below are some guidelines for proper backfilling procedures for PVC fittings:

1. If rocky conditions exist at the trench bottom, a minimum layer of 4 in. (100 mm) of bedding should be placed below the fitting. If the trench bottom is well compacted, then bell holes should be excavated to ensure uniform loading on the fitting. Sand bedding loosely placed would not require bell hole excavation.

2. A quality backfill material should then be placed and compacted around the sides of the fitting and to a level of 6 in. (150 mm) above the top of the fitting. Material should be compacted using only tamping bars when within 6 in. (150 mm) of the fitting. Jumping jacks or plate tampers may be used elsewhere.

3. Backfill material for the bedding, haunch, and initial backfill zones must contain stones no larger than 1½ in. (38 mm) in size.

4. Final backfill should be free of sharp objects, sticks, or boulders. Hoe-pacs or vibratory rollers should never be used within 3 ft (1 m) (after compaction) of PVC pipe or fittings.

5. Compaction is especially important on both sides of a branch of a PVC tee or wye to minimize lateral bending stresses.

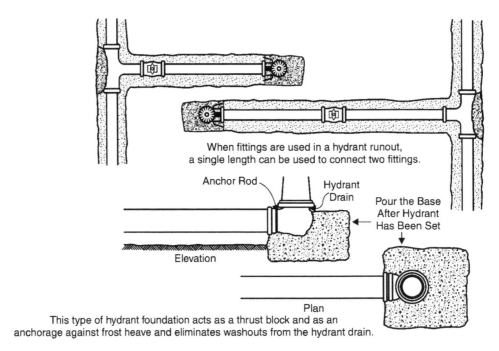

Source: Courtesy of Johns-Manville Sales Corp.

Figure 7-6 Fire hydrant foundation

Joint Restraint

Thrust-block requirements for PVC fittings do not differ from those presented earlier for other fittings. Thrust developed will depend upon the configuration and diameter of the fitting as well as the system internal pressure. It is recommended that a layer of plastic sheeting be placed over the fitting before pouring a concrete thrust block to achieve easy removal if required in the future. It is also recommended that precast concrete thrust blocks not be used with PVC fittings because a point loading may be inflicted on the wall of the fitting.

Joint restraint may also be obtained in unstable soils or common trenches by using mechanical restraint devices. These devices have been used successfully on PVC pressure fittings up to 36 in. (900 mm) diameter. Devices should conform to the requirements of ASTM F1674.

Installation

The full weight of valves, hydrants, and fittings should not be carried by the pipe. Valves, hydrants, and fittings should be provided with individual support, such as treated timbers, crushed stone, concrete pads, or a well-compacted trench bottom. Valves should connect directly to PVC pipe using elastomeric gaskets supplied by the valve manufacturers.

Control valves. Control valves in a pressurized system should be checked to ensure that sufficient resistance to thrust is available when the valve is closed. In some designs, butterfly valves will not function properly on certain sizes of PVC pipe without special nipple adapters.

Fire hydrants. All fire hydrants shall stand plumb, shall be properly located and oriented, and shall be set to proper elevation. The constructor shall provide a coarse-aggregate drain pocket or drain pit for dry-barrel hydrants. The installation recommendations of AWWA M17, *Installation, Field Testing, and Maintenance of Fire Hydrants,* shall be followed. (See Figure 7-6, Fire Hydrant Foundation.)

INSTALLATION 91

Fittings. Some fittings in pressurized systems require reaction or thrust blocking to prevent movement caused by longitudinal line thrust.

THRUST RESTRAINT

Water under pressure exerts thrust forces in a piping system. Thrust blocking or joint restraint should be provided to prevent movement of pipe or appurtenances in response to thrust. Thrust blocking or joint restraint, which conforms to UNI-B-13 *Performance Specification for Joint Restraint Devices for Use with Polyvinyl Chloride (PVC) Pipe*, are required wherever the pipeline

- changes direction (e.g., tees, bends, elbows, and crosses)
- changes size, as at reducers
- stops, as at dead ends
- valves and hydrants are closed and thrust develops

Size and type of thrust blocking or joint restraint depends on the following:

- Maximum system pressure (including field testing pressures)
- Mipe size
- Appurtenance size
- Type of fittings or appurtenance
- Line profile (e.g., horizontal or vertical bends)
- Soil type

Figures 7-7 and 7-8 display standard types of thrust blocking and joint restraint used in pressurized water systems. Thrust restraint and joint restraint design are discussed in chapter 4, starting with the Thrust Restraint-General section.

The mechanical joint shall be assembled in accordance with the fitting manufacturer's recommendations. Pipe spigot bevels may require shortening for use with mechanical joints or fittings joints.

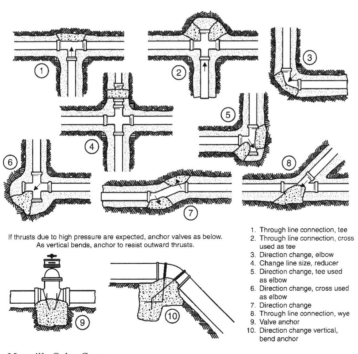

If thrusts due to high pressure are expected, anchor valves as below. As vertical bends, anchor to resist outward thrusts.

1. Through line connection, tee
2. Through line connection, cross used as tee
3. Direction change, elbow
4. Change line size, reducer
5. Direction change, tee used as elbow
6. Direction change, cross used as elbow
7. Direction change
8. Through line connection, wye
9. Valve anchor
10. Direction change vertical, bend anchor

Source: Courtesy of Johns-Manville Sales Corp.

Figure 7-7 Types of thrust blocking

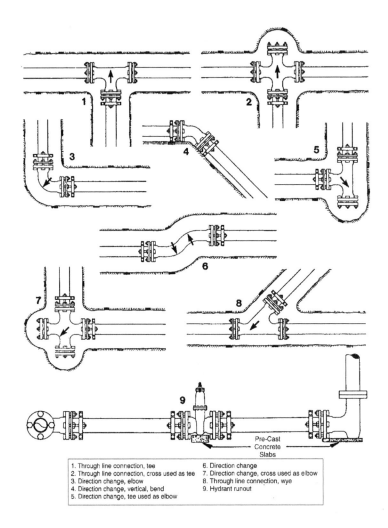

Figure 7-8 Types of joint restraint

AWWA MANUAL M23

Chapter **8**

Testing and Maintenance

This chapter contains procedures for leakage and pressure testing and for disinfection before PVC or PVCO potable water piping is placed in service. Procedures are also presented for thawing and locating PVC and PVCO piping. Installers should refer to the contract documents for the required testing procedures for new construction.

INITIAL TESTING

During the initial stages of the construction project, a short length of installed pipeline should be tested. This will permit the installer to verify that proper installation and joint assembly techniques have been employed.

The length of section to be tested will depend on the specific project; however, the length should be long enough to ensure a representative length is used with at least 20 pipe joints. To avoid additional costs caused by restrained joints and temporary blocking, it may be possible to adjust the initial test length to suit facilities already required for final installation.

The requirement for initial testing should be clearly stated in the contract documents to avoid a claim for delay of work. Some contractors may fail to see that the requirement for initial testing is beneficial to all involved in the project. The initial test pressure should be to the same hydraulic grade line as the final testing. The testing procedure is presented later in this chapter.

TIMING OF THE TESTING

To prevent floating of the pipe, sufficient backfill should be placed prior to filling the pipe with water and subsequent field testing. Where local conditions require that the trenches be backfilled immediately after the pipe has been laid, the testing may be completed after backfilling but before placement of the permanent surface.

At least seven days should elapse after the last concrete thrust or reaction blocking has been cast with normal (Type 1) portland cement. The elapsed time may be reduced to three days with the use of a high-early-strength (Type III) portland cement.

Restrained joints must be assessed if testing is to be done immediately after trench backfilling. At this stage of installation, the full weight of the trench backfill may not have come to bear on the pipe. If this is the case, the frictional resistance at the time of the testing may be less than that anticipated by the designer. The designer of the water main should be consulted prior to the testing.

INITIAL CLEANING OF THE PIPELINE

The pipe must be cleaned before disinfecting the water main. Mud, sand, dirty water, or a variety of debris left in the water main during construction will shield bacteria from contact with the chlorine solution resulting in incomplete disinfection and possibly delivery of contaminated water to consumers. Water mains may be successfully cleaned by flushing or swabbing.

Water mains should be filled slowly with potable water at a rate that allows air to leave the line at the same rate as the water entering the line. (Example: 2-in. air valve, $Q = 2.18$ cubic feet per second [ft^3/sec].) All air should be vented from the pipe and fittings to prevent entrapment of air in the main. For flushing of the pipeline, it is suggested that sufficient flow in the system be created to cause the pipeline flow velocity to be equal to or greater than 3.0 ft/sec. The duration of the initial flushing procedure should be continued until the discharge appears clean; however, the minimum duration should be based on a minimum of three changes of pipeline volume. The discharge of flushing water must be carefully managed to prevent damage to property or interruption to traffic.

Foam swabbing may be used in lieu of flushing. This procedure should be carefully assessed prior to undertaking the work.

TEST PREPARATION

Proper preparation for the hydrostatic and leakage tests will help ensure that the test is conducted in a safe and efficient manner. It is critical that all facilities required to handle thrust restraints are properly installed. Thrust facilities must also include the trench backfill if restrained joints are used. Concrete thrust blocks must be allowed to develop the required strength.

To reduce the amount of make-up water to its lowest level and make the test more meaningful, the following procedures should be performed:

1. Close all outlets.
2. Tighten all bolts on water main appurtenances and flanges.
3. Release air from **all** high points.
4. Fill the main with test water from a low point.
5. Fill the line at a slow rate (maximum velocity of 1 ft/sec [0.3 m/sec]) to minimize air entrapment and potential surge pressures.
6. After filling the line, flush the line for a sufficient period to help move the air to the release valves or other outlets.
7. Check the operation of all pumping equipment and calibrate or certify gauges and meters prior to making test.

HYDROSTATIC TESTING AND LEAKAGE TESTING

Hydrostatic testing is required to prove the integrity of the pipeline. The tests are usually conducted by the contractor and are witnessed by the owner or the owner's representative. Hydrostatic testing is usually done over a short duration, usually 1 hr. The hydrostatic test is done initially to determine if the pipeline was put together properly and may be combined with the leakage test.

The leakage test is similar to the hydrostatic test except that the amount of make-up water required to maintain a given pressure is measured. If the quantity of make-up water added is less than a predetermined value, the pipeline is considered acceptable.

In addition to leakage during testing, there are several additional causes that may lead to the requirement to add make-up water. These factors include compressing trapped air, take-up of restraints, and temperature variations during testing. All visible leaks observed during the leakage test should be repaired. The contract documents should include an allowance for leakage; however, some authorities state that no leakage should be permitted.

TEST PRESSURE

To test the system, the hydrostatic pressure should be increased to at least the maximum operating pressure of the line. Many jurisdictions require that the system be subjected to pressures in the range of 120–150 percent of the maximum working pressure. The allowance over the maximum working pressure should be carefully considered. A high allowance can increase the cost of the system as a result of the size of thrust blocks, number of restrained joints, and the increased pressure rating of appurtenances such as flanges, valves, etc.

The required hydrostatic test pressure should be presented as a hydraulic grade line rather than a gauge pressure. This is particularly important in hilly terrain. A vertical rise of 100 ft (33 m) translates into approximately 43 psi (296 kPa). Thus, if the hydrostatic test pressure is identified as 150 psi (1,034 kPa), then to get a reading of 150 psi (1,034 kPa) at the top of a 100-ft-high (33-m) hill, the pressure would be 193 psi (1,330 kPa) at the bottom of the hill. Stating the test pressure in terms of hydraulic grade line makes it easier to understand the effects of the test pressure on the pipeline design.

Prior to establishing the required test pressure, the designer must consider the following operating conditions of the system:

- Normal working pressure
- Maximum sustained operating pressure
- Maximum transient pressure along the pipeline

As an example, consider a water main installed in a system with the following operating conditions at the most critical location:

- Working pressure–60 psi
- Maximum sustained operating pressure–85 psi
- Maximum transient pressure–110 psi

In this case, a test pressure of 132 psi would provide a test pressure ratio of 1:20 on the maximum transient pressure, 1:55 on the maximum operating pressure, and 2:20 on the working pressure. If the water main elevation is 202 ft and the elevation of a hydrant at the critical location is 208 ft, the required hydraulic grade line for the test is 507 ft (132 psi = [507 −202] × .43). Therefore, the required gauge pressure as measured at the hydrant is 129 psi.

DURATION OF TESTS

The duration of the tests should be specified in the contract documents. Leakage tests should be conducted over periods long enough to determine the average leakage rates. It is suggested that a separate hydrostatic pressure test should last 1 hr. A separate leakage test should be conducted over a 2-hr period. If the leakage test is conducted simultaneously with hydrostatic pressure test, the test should be conducted over a 2-hr period.

ALLOWABLE LEAKAGE

The allowable leakage rate should be identified in the contract documents. In addition to any leakage allowance, the system should be examined under pressure and all visible leaks shall be repaired regardless of the amount of leakage.

It is suggested that no installation will be accepted if the leakage is greater than that determined by the formula:

$$L = \frac{ND\sqrt{P}}{7,400} \tag{8-1}$$

Where:

L = allowable leakage, gph
N = number of joints in the length of pipeline to be tested
D = nominal diameter of the pipe, in.
P = test pressure, psi

This leakage rate is based on 10.5 gpd/mi/in. of nominal diameter at a pressure of 150 psi. Table 8-1 summarizes allowable leakage for a range of test pressures and pipe diameters.

Do not test the water main in sections that are too long. With long test sections, large leaks can go undetected because of the averaging effect. The designer must consider placement of isolation valves with testing in mind to accommodate reasonable test sections.

DISINFECTING WATER MAINS

All potable water mains must be disinfected prior to being put into service. This very important procedure should be done under the direct control of the operating authority or its agents. Disinfection should comply with AWWA C651.

The disinfection should take place after the initial cleaning and after the completion of the hydrostatic and leakage testing. Water from the existing water distribution system is allowed to flow into the section of water main to be disinfected at a controlled rate.

All chlorinated water used for testing, flushing, or disinfecting water mains should be disposed of safely. High chlorine residuals may cause significant negative health effects on wildlife and humans. Confirm acceptable levels of chlorine with local, state, or federal authorities.

If the chlorine residual in the water main under disinfection remains relatively high, then usually there are no impurities in the systems creating a chlorine demand. However, if the chlorine applied to the section is gone, it can be concluded there are impurities in the system.

After the water main has been recharged after successful chlorination, samples should be taken for bacteriological tests according to C651 24 hr after refilling.

The system must not be put into service until the operating authority or its agents are satisfied with the testing procedure.

SYSTEM MAINTENANCE

For PVC water mains, as well as those made from other materials, it is important to set up a system maintenance program. A good maintenance program accomplishes the following:

- Helps prevent failure of facilities
- Detects problems in the water systems

Table 8-1 Allowable leakage per 50 joints of PVC pipe,* gph†

		Nominal Pipe Diameter—in. (mm)													
Avg. Test Pressure, psi (kPa)	4 (100)	6 (150)	8 (200)	10 (250)	12 (300)	14 (350)	16 (400)	18 (450)	20 (500)	24 (610)	30 (760)	36 (915)	42 (1,050)	48 (1,200)
300 (2,070)	0.47	0.70	0.94	1.17	1.40	1.64	1.87	2.11	2.34	2.81	3.51	4.21	4.92	5.62
275 (1,900)	0.45	0.67	0.90	1.12	1.34	1.57	1.79	2.02	2.24	2.69	3.36	4.03	4.71	5.38
250 (1,720)	0.43	0.64	0.85	1.07	1.28	1.50	1.71	1.92	2.14	2.56	3.21	3.85	4.49	5.13
225 (1,550)	0.41	0.61	0.81	1.01	1.22	1.42	1.62	1.82	2.03	2.43	3.04	3.65	4.26	4.86
200 (1,380)	0.38	0.57	0.76	0.96	1.15	1.34	1.53	1.72	1.91	2.29	2.87	3.44	4.01	4.59
175 (1,210)	0.36	0.54	0.72	0.89	1.07	1.25	1.43	1.61	1.79	2.15	2.68	3.22	3.75	4.29
150 (1,030)	0.33	0.50	0.66	0.83	0.99	1.16	1.32	1.49	1.66	1.99	2.48	2.98	3.48	3.97
125 (860)	0.30	0.45	0.60	0.76	0.91	1.06	1.21	1.36	1.51	1.81	2.27	2.72	3.17	3.63
100 (690)	0.27	0.41	0.54	0.68	0.81	0.95	1.08	1.22	1.35	1.62	2.03	2.43	2.84	3.24
75 (520)	0.23	0.35	0.47	0.59	0.70	0.82	0.94	1.05	1.17	1.40	1.76	2.11	2.46	2.81
50 (340)	0.19	0.29	0.38	0.48	0.57	0.67	0.76	0.86	0.96	1.15	1.43	1.72	2.01	2.29

* If the pipeline under test contains sections of various diameters, the allowable leakage will be the sum of the computed leakage for each size.
† To obtain leakage in litres per hour, multiply the values in the table by 3.72.

TESTING AND MAINTENANCE

- Determines the necessary replacement parts
- Provides input for future installation
- Maintains good public relations
- Detects and eliminates safety hazards
- Distributes maintenance work more advantageously
- Reduces cost of system maintenance

Records should be kept of all components of the work systems including date of installation, materials, size, pressure class, operating pressures, soil conditions, installer's name, and any other relevant information.

Records should be kept of pipe breakage, leak surveys, pressure tests, and friction coefficient tests to assist in decision making relating to maintenance and replacement of water distribution facilities.

The high dielectric strength of PVC pipe prevents the location of buried PVC pipe with electric current type metal detectors unless a tracer has been buried with the pipe during installation. If location during maintenance and future construction is to be accomplished using these types of detectors, a low-cost inductive or conductive tracer should be buried with the pipe (for example, metallic ribbon or other metal wire). Location also may be accomplished using a buried pipe detection apparatus designed to locate nonmetallic buried pipelines (such as asbestos-cement and plastic). These devices operate in a manner similar to a simple seismograph and have proven to be effective. Probing with metal bars is not recommended.

Frozen PVC water lines may be thawed by hot-water or steam injection. Torches and other direct-heating devices should not be used to thaw frozen lines.

AWWA MANUAL M23

Chapter 9

Service Connections

Service connections vary in size from small services supplying individual homes to large outlets for industrial users. Service lines can be connected to PVC water mains while either in or out of service using the following methods:

- Direct tapping (6 in. through 12 in. [150 mm through 300 mm] AWWA C900, PC 150 and PC 200 only, up to 1 in. [25 mm] corporation stops); PVCO pipe cannot be direct tapped
- Tapping through service saddles (up to 2 in. [50 mm] corporation stops)
- Tapping sleeves and valves (up to size-to-size)

In all cases, service lines are to be properly backfilled to provide support independent of the corporation stop or pipe.

DIRECT TAPPING

General

Direct tapping involves tapping threads into the pipe wall and inserting a corporation stop (see Figure 9-1). Direct tapping is recommended for PVC water distribution pipe manufactured in accordance with AWWA C900 in nominal sizes 6 in. through 12 in. (150 mm through 305 mm), Pressure Classes 150 and 200.

Direct tapping is not recommended for Pressure Class 100 pipe, 4-in. Pressure Classes 150 and 200 pipe, AWWA C905 pipe, or AWWA C909 pipe. In these cases, service clamps or saddles should be used.

Equipment

Tapping machine. A tapping machine is a mechanical device used to install the direct tapped connection into water mains and may vary in design and operation depending on the specific machine manufacturer. The machine must operate with a cutting/tapping tool that is classified as a core cutting tool (either with internal teeth or with double slots) of the shell design and which retains the coupon cut while penetrating the wall of the water main. The tapping machine shall provide the standard

Figure 9-1   Direct PVC pipe tap

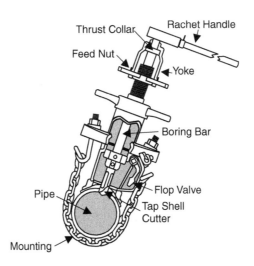

Figure 9-2   Tapping machine nomenclature

ratchet handle on the boring bar as indicated in Figure 9-2. The tapping machine shall be of a design where cutting and tapping is controlled and accomplished with a feed nut or feed screw and yoke.

Cutting/tapping tools. Figure 9-3 shows a combination drill and tap tool which illustrates the recommended features of the cutter. A core drill is essential and may be either plain or slotted. Slotted core drills allow an easier cut, but not a faster cut. The cutter must be sharp and maintained throughout its service life to ease the cut and thereby greatly increase the success of the operation. Dull cutters can lead to "push through" at the inside diameter penetration, which may cause failure. **Caution: Do not drill a hole in PVC pipe with a twist drill or auger bit.**

The core drill must retain the plug of material or coupon removed from the pipe wall. A simple means of removing the coupon from the core drill should be provided. The core drill must be designed to accommodate walls as heavy as DR 14 (Pressure Class 200, AWWA C900). Sufficient throat depth is required. The shank of the cutter must be adaptable to the tapping machine being used.

Figure 9-3   Cutting/tapping tool

The tap must cut AWWA C800-tapered threads. Iron pipe threads are not recommended for the pipe wall. The depth of travel of the cutter as the threads are cut must be carefully adjusted (to ensure the proper insertion of the corporation stop with 2–3 threads showing after applying an insertion torque of 27–35 ft-lb [36.6–47.5 N-m]).

Corporation stop. The corporation stops should be AWWA-tapered with thread complying with AWWA C800 in sizes ⅝ in. (16 mm), ¾ in. (19 mm), and 1 in. (25 mm). When sizes larger than 1 in. (25 mm) are required, tapping saddles or sleeves should be used.

Direct Tapping Procedure

When planning a direct tap, the following should be considered:

- Only AWWA C900 Pressure Class 150 and 200 PVC pipes 6 in. (150 mm) through 12 in. (300 mm) can be direct tapped.
- Taps up to 1 in. (25 mm) can be made directly (i.e., ⅝ in. [16 mm], ¾ in. [19 mm], and 1 in. [25 mm]).
- Wet taps or dry taps (pipe filled or pipe empty) are allowed.
- The maximum allowable pressure in the pipe at the time the wet tap is being completed is the rated pressure class of the pipe (i.e., DR 18, PC 150; DR 14, PC 200).
- Corporation stops must have AWWA C800 thread. Iron pipe threads are not recommended for the pipe wall.
- Use a combination core drill and tap when tapping direct. Do not use twist drills designed for tapping hardwall pipes.
- Recommended temperature limits as follows:
    —Dry taps: 0°F (–18°C) to 100°F (38°C)
    —Wet taps: 32°F (0°C) to 90°F (32°C)

These temperatures refer to the temperature of the pipe itself, which is closely related to air temperature in most cases.

- Placement
    —Tap no closer than 2 ft (600 mm) from the ends of the pipe.
    —Stagger multiple taps and keep them at least 18 in. (450 mm) apart lengthwise.
    —Avoid tapping into a discolored surface.
    —Do not tap a curved pipe if the radius of the bend is less than 300 times the pipe outside diameter.
- Do a test tap in the shop before starting field tapping by performing the following:
    —Try a couple of dry bench taps to establish the mark on the boring bar corresponding to the correct tapping depth.
    —Firmly seat and secure the cutter in the holder of the tapping machine. Secure it well; any wobbling or looseness of either the cutter or the boring bar will cause problems.

Table 9-1 PVC pipe outside diameters

Nominal Pipe Size		Pipe OD,
in.	(mm)	in.
6	(150)	6.90
8	(200)	9.05
10	(250)	11.10
12	(300)	13.20

— Make a final check of thread compatibility. AWWA threads are required for both the tap and the corporation stop.

- Observe the following basic tapping precautions when tapping pressurized pipes:
    — Have a second person close by.
    — Wear protective goggles.
    — Provide a quick exit.
    — Cover the pipe area with a protective blanket (without obstructing machine operation).
    — Follow local regulations.

When tapping pressurized pipes, the personnel on the surface should have a clear understanding of the valve operation necessary to isolate the tapping site, if required.

Mounting the machine. Make sure the outside diameter (OD) of the pipe falls within the range of ODs for which the tapping machine saddle is designed (Table 9-1).

The machine should sit on the drilling site firmly, but not in a way that will set up wall stresses by distorting the pipe. The tapping machine should be placed on the pipe in accordance with the recommendations of the machine manufacturer.

When taps are being made on the horizontal plane (the preferred location because it keeps the gooseneck of the service pipe as far below the frostline as possible), it is important that the tightening nuts be turned down evenly on each side. The following procedure should be used:

1. Adjust the nuts on the chain hooks so that they are even with the tops of the threads.
2. Position the chain hooks on the machine and loop the chain links into the hooks snugly.
3. Tighten nuts A and B alternately so that the same number of threads are showing when the machine is correctly and firmly mounted (see Figure 9-4).
4. During the process, make sure the machine remains correctly seated in the saddle and saddle gasket. Overtightening on only one side may distort the wall and set up wall stresses. Wrench extenders should not be used.

Cutting the hole and tapping the threads. Because PVC is easy to machine, it is tempting to overfeed the cutter because it is comparatively easy to turn the ratchet handle. The cutters should be fed lightly—just enough to keep the cutter engaged. The principle is to allow the cutter to work as a cutter. Under normal conditions, the following procedure should be used:

1. Rotate the ratchet handle one complete turn for every ⅛ turn of the feed yoke (approximately 1-in. [25 mm] movement of feed nut to each ¼-turn of ratchet handle) (see Figure 9-5). The feed rate should be less in cold weather. Judge the correct feed rate by "finger pull"—the effort should be

SERVICE CONNECTIONS 103

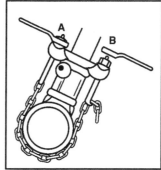

Wrong — Overtightening only one side may distort wall and stress pipe. Never use wrench extensions.

Right — Both sides evenly tightened using only the wrench supplied.

Figure 9-4 Mounting the tapping machine

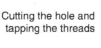
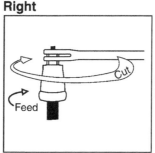

Right — Cutting the hole and tapping the threads. Feed lightly—just enough to keep the cutter engaged.

Figure 9-5 Cutter feed

about the same used to open a desk drawer. This rule of thumb applies in any temperature.

2. Upon wall penetration, the upward thrust on the boring bar (assuming a pressurized pipe) will be about 1 lb-force per 1 psi of water pressure.

3. Use the feed yoke to engage the first few turns of the tapping tool in the hole. After this, the tap is self-feeding and the feed yoke can be disengaged from the boring bar.

4. The "cast-iron" mark on the boring bar is not a reliable indicator of how deep to tap. Tapping to the correct depth is important and should be determined by performing bench tests in advance and carefully noting the position of the top of the threaded feed sleeve, relative to the thrust collar or other datum point, when the corporation stop is correctly inserted.

5. As the tapping tool is reversed out of the hole, re-engage the feed yoke or hold the boring bar until the tap clears the threads. Release the bar slowly so as not to damage the threads or injure the machine operator.

6. Examine the coupon of PVC after it is knocked out of the cutter head. A clean edge means good cutting action. A raised crown means the cutter was fed through too fast. If the cutter is fed too fast, a plug of material is pushed out of the ID of the pipe where the tool emerges. The condition of the coupon

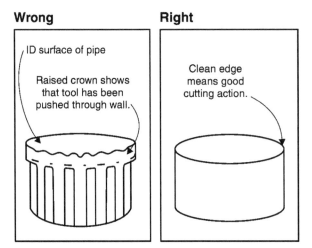

Figure 9-6 Condition of coupon

provides a check of correct tapping procedure (see Figure 9-6). At the first sign of a crown on the coupons, the tapping procedure or the condition of the tools should be reexamined and corrected before more taps are attempted.

Inserting the stop. The following should be performed to insert a stop:

1. Attach the E-Z screw plug to the end of the boring bar. Screw the corporation stop into the E-Z release plug. The exposed end of the main stop will be the inlet end with tapered (AWWA) threads. Check to BE SURE THE STOP IS CLOSED.

2. Apply two spiral wraps of Teflon tape clockwise to the AWWA threads. Other thread lubricants are not recommended. Do not use liquid sealants, even though they may contain Teflon.

3. Replace the boring bar assembly in the machine and insert the stop into the main to start the first few threads in the hole so that they are not forced or punched. Using the feed yoke for this operation requires only light finger pressure on the feed nut while the ratchet handle is rotated.

4. Disengage the feed yoke and remove the ratchet handle as soon as the stop has firmly engaged the threads in the pipe wall. Complete the insertion using a torque wrench.

5. Tighten the stop to 27 ft-lb (36.6 N-m).

6. Snap the wrench counterclockwise to release the E-Z plug from the stop. Remove the tapping machine. If there is leaking past the threads, tighten the stop to 35 ft-lb (47.5 N-m). At correct insertion, one to three threads should be visible.

7. If leaking past the threads persists, do the following:
   a. Remove pressure from the pipe.
   b. Unscrew the stop.
   c. Clear away cuttings.
   d. Remove old tape and replace with new Teflon tape.
   e. Replace the stop and tighten to 27 ft-lb (36.6 N-m).

Direct Dry Tapping

When direct tapping AWWA C900 DR 18 Pressure Class 150 or DR 14 Pressure Class 200 PVC pipe 6 in. (150 mm) through 12 in. (300 mm) that is empty, or not in service, or not yet under pressure, the procedure given for making wet taps under pressure should be followed, with a few exceptions:

1. Remove the machine from the pipe after the hole has been drilled and tapped. Carefully remove the cuttings from the hole before inserting the main stop.

2. Prepare the stop as described in the section on inserting the stop and insert by hand. **Caution: Take special care not to crossthread the fine Teflon-wrapped AWWA threads.**

3. Tighten the stop to the point where one to three threads are showing at 27 ft-lb (36.6 N-m).

SADDLE TAPPING

General

The use of saddles (see Figures 9-7 and 9-8) to make taps in PVC pressure pipe is recommended for any size or class of pipe. Service connections may be made using a service clamp or saddle. Maximum outlet size recommended with service clamps or saddles is 2 in. (50 mm). Operating or in situ conditions such as severely expansive native soils may also encourage saddle selection. When making this type of connection, use equipment that attaches to the corporation stop permitting a cutting tool to be fed through the corporation stop to cut a hole in the pipe. Tapping of the pipe wall is not required since the corporation stop is threaded into the service clamp.

Equipment

Service clamps or saddles. Service clamps or saddles used for attaching service connections to PVC water pipe should

- provide full support around the circumference of the pipe
- provide a bearing area of sufficient width along the axis of the pipe, 2 in. (50 mm) minimum, ensuring that the pipe will not be distorted when the saddle is tightened

Service clamps should **not** have

—lugs that will dig into the pipe when the saddle is tightened

—a U-bolt type of strap that does not provide sufficient bearing area

—a clamping arrangement that is not fully contoured to the outside diameter of the pipe

Tapping machine. A number of tapping machines are available that will cut through a corporation stop. The tapping machine may vary in design and operation depending on the specific machine manufacturer. The machine must operate with a cutting tool that is classified as a core cutting tool (either with internal teeth or with double slots) of the shell design, which retains the coupon cut while penetrating the wall of the pipe. The tapping machine shall have the standard ratchet handle on the boring bar. The tapping machine shall be of a design where cutting is controlled and accomplished with a feed nut or feed screw and yoke.

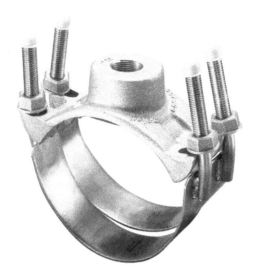

Source: Romac Industries.

Figure 9-7 Tapping saddle

Cutting tool. It is important that the cutting tool be a shell type (hole) cutter, which will retain the coupon and be designed to accommodate walls as heavy as DR 14 (Pressure Class 200, AWWA C900). Many shell cutters are designed only for thin-walled PVC, consequently, they do not have sufficient throat depth to handle the heavier walled pipe. **Caution: Do not drill a hole in PVC pipe with a twist drill or auger bit.** The shank of the cutter must be adaptable to the cutting machine being used.

Corporation stop. Because the corporation stop is inserted into the service clamp or saddle, it must have threads that match that of the clamp or saddle, which may be either IPS-OD or CIOD. The maximum size of corporation stop which may be used with a service clamp or saddle is 2 in. (50 mm). If a tap larger than 2 in. (50 mm) is required, a tapping sleeve and valve are required.

Saddle Tapping Procedure

1. Evenly tighten the saddle to the pipe. Screw the inlet side of the corporation stop into the saddle threads. Retain the stop hardware as supplied.
2. Open the corporation stop.
3. Using the appropriate adapter and gasket, attach the drilling machine to the main stop outlet threads. Use a machine with an operator-controlled feed rate. The use of a core drill is essential. Follow the machine manufacturer's instructions.
4. Lower the boring bar to the main and rotate the cutter while exerting finger-pull pressure on the feed handle. The principle is to allow the cutter to work as a **cutter**. Rotate the ratchet handle one complete turn for every $\frac{1}{8}$-turn of the feed yoke (approximately 1-in. [25-mm] movement of feed nut to each $\frac{1}{4}$-turn of ratchet handle).
5. Withdraw the cutter, close the corporation stop, and remove the drilling machine.

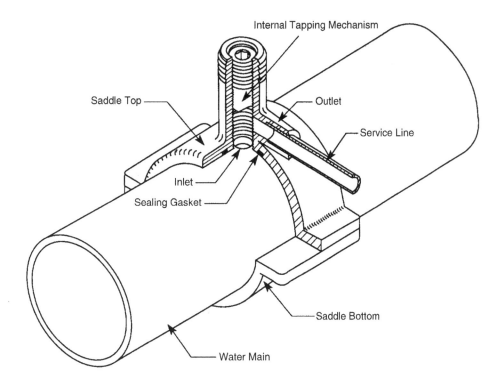

Figure 9-8 PVC tapping saddle

TAPPING SLEEVE AND VALVE

General

Tapping sleeves and valves are used when service connections larger than 2 in. (50 mm) must be made in PVC pipes. Tapping sleeves may be used for making larger taps under pressure. The use of tapping sleeves and valves to make taps in PVC pressure pipe can be recommended for any size or class of pipe. When making this type of connection, equipment is used which attaches to the valve permitting a cutting tool to be fed through the valve to cut a hole in the pipe. Tapping of the pipe wall is not required because the valve is attached to the tapping sleeve.

Equipment

Tapping sleeves and valves. When tapping sleeves are ordered from the manufacturer, specify the outside diameter of the pipe being tapped, the size of the outlet, and the working pressure of the system. Lead-joint sleeves should not be used.

Tapping sleeves should

- provide full support around the circumference of the pipe
- provide a minimum bearing (laying) length (consult tapping sleeve manufacturer)

Tapping sleeves should not

- distort the pipe when tightened
- have lugs that will dig into the pipe when the sleeve is tightened
- have a clamping arrangement which is not fully contoured to the outside diameter of the pipe

When ordering tapping valves, specify dimensions and attaching mechanisms consistent with the tapping sleeves.

Cutting tool. A toothed core cutter that retains the coupon, similar to those used for other materials, should be used. The cutter should have sufficient throat depth to cut heavy-walled PVC pipe such as DR 14. **Caution: Do not drill a hole in PVC pipe with a twist drill or auger bit.**

Tapping machine. A number of tapping machines are available that will cut through a tapping valve. The tapping machine may vary in design and operation depending on the specific machine manufacturer. The tapping machine must be attached to the valve and must be ordered according to the valve specified. Tapping equipment can be purchased or rented from sleeve manufacturers who also furnish instructions and/or instructors trained in making such taps. (Contractors who specialize in this type of work are also available in some areas.)

Tapping Sleeve and Valve Procedure

The tapping sleeve should be assembled on the pipe in accordance with the manufacturer's directions, ensuring that no pipe distortion occurs. The tapping valve is then connected to the sleeve.

Tapping sleeves should be supported independently from the pipe during the tapping. Support used should be left in place after tapping. Thrust blocks should be used as with any other fitting or appurtenance.

Attach the drilling machine and adapter to the valve outlet flange. Position the necessary support blocks. Open the tapping valve, advance the cutter, and cut the hole into the pipe through the sleeve. Retract the cutter and then close the tapping valve. Remove the drilling machine and attach the new pipe.

This page intentionally blank.

AWWA MANUAL M23

Appendix A

Chemical Resistance Tables

Table A-1 Chemical resistance of PVC pressure water pipe

Chemical	73°F (23°C)	140°F (60°C)	Chemical	73°F (23°C)	140°F (60°C)
Acetaldehyde	N	N	Ammonia, liquid	N	N
Acetaldehyde, aq 40%			Ammonium fluoride, 25%	R	C
Acetamide			Ammonium salts, except fluoride	R	R
Acetic acid, 20%			Amyl acetate	N	N
Acetic acid, 80%			Amyl chloride	N	N
Acetic acid, glacial			Aniline	N	N
Acetic acid, vapor			Aniline chlorohydrate	N	N
Acetic anhydride	C	N	Aniline dyes	N	N
Acetone	—	—	Aniline hydrochloride	N	N
Acetylene	R	R	Anthraquinone	R	R
Adipic acid	R	C	Anthraquinone sulfonic acid	R	R
Alcohol, allyl	R	N	Antimony trichloride	R	R
Alcohol, benzyl	R	R	Aqua regia	C	N
Alcohol, butyl (2-butanol)	N	N	Arsenic acid, 80%	R	R
Alcohol, butyl (n-butanol)	N	N	Aryl-sulfonic acid	R	R
Alcohol, ethyl	C	C	Barium salts	R	R
Alcohol, hexyl	R	R	Beer	R	R
Alcohol, isopropyl (2-propanol)	R	C	Beet sugar liquor	N	R
Alcohol, methyl	N	N	Benzaldehyde, 10%	R	N
Alcohol, propyl (1-propanol)	R	R	Benzaldehyde, above 10%	N	N
Allyl chloride	N	N	Benzene (benzol)	N	N
Alums	R	R	Benzene sulfonic acid, 10%	R	R
Ammonia, aq	R	R	Benzene sulfonic acid, above 10%	N	N
Ammonia, gas	R	R	Benzoic acid	R	R

(continued on next page)

109

Table A-1 Chemical resistance of PVC pressure water pipe (continued)

Chemical	73°F (23°C)	140°F (60°C)	Chemical	73°F (23°C)	140°F (60°C)
Black liquor-paper	R	R	Chlorobenzyl chloride	N	N
Bleach, 5.5% active chlorine	R	R	Chloroform	N	N
Bleach, 12.5% active chlorine	R	R	Chlorosulfonic acid	R	N
Borax	R	R	Chromic acid, 10%	R	R
Boric acid	R	R	Chromic acid, 30%	R	C
Boron trifluoride	R	R	Chromic acid, 40%	R	C
Bromic acid	R	R	Chromic acid, 50%	N	N
Bromine, aq	R	R	Citric acid	R	R
Bromine, gas, 25%	R	R	Coconut oil	R	R
Bromine, liquid	N	N	Coke oven gas	R	R
Butadiene	R	R	Copper salts, aq	R	R
Butane	R	R	Corn oil	R	R
Butanediol	R	R	Corn syrup	R	R
Butantetrol (erythritol)	R	N	Cottonseed oil	R	R
Butyl acetate	N	N	Cresol	N	N
Butyl phenol	R	N	Cresylic acid, 50%	R	R
Butylene	R	R	Croton aldehyde	N	N
Butyric acid	R	N	Crude oil	R	R
Calcium hydroxide	R	R	Cyclohexane	N	N
Calcium hypochlorite	R	R	Cyclohexanol	N	N
Calcium salts, aq	R	R	Cyclohexanone	N	N
Can sugar liquors	R	R	Detergents, aq	R	R
Carbon bisulfide	N	N	Diazo salts	R	R
Carbon dioxide	R	R	Dibutyl phthalate	N	N
Carbon dioxide, aq	R	R	Dibutyl sebacate	N	N
Carbon monoxide	R	R	Dichloroethylene	N	N
Carbon tetrachloride	R	N	Diesel fuels	R	R
Casein	R	R	Diethyl amine	N	N
Castor oil	R	R	Diglycolic acid	R	R
Caustic potash (potassium hydroxide)	R	R	Dimethyl formamide	N	N
Caustic soda (sodium hydroxide)	R	R	Dimethylamine	R	R
Cellosolve	R	C	Dioctyl phthalate	N	N
Cellosolve acetate	R	—	Dioxane-1,4	N	N
Chloracetic acid	R	R	Disodium phosphate	R	R
Chloral hydrate	R	R	Ethers	N	N
Chloramine	R	—	Ethyl esters	N	N
Chloric acid, 20%	R	R	Ethyl halides	N	N
Chlorine, gas, dry	C	N	Ethylene glycol	R	R
Chlorine, gas, wet	N	N	Ethylene halides	N	N
Chlorine, liquid	N	N	Ethylene oxide	N	N
Chlorine water	R	R	Fatty acids	R	R
Chlorobenzene	N	N	Ferric salts	R	R

(continued on next page)

Table A-1 Chemical resistance of PVC pressure water pipe (continued)

Chemical	73°F (23°C)	140°F (60°C)	Chemical	73°F (23°C)	140°F (60°C)
Fluoboric acid, 25%	R	R	Jet fuels, JP-4 and JP-5	R	R
Fluorine, dry gas	C	N	Kerosene	R	R
Fluorine, wet gas	C	N	Ketones	N	N
Fluosilicic acid	R	R	Kraft paper liquor	R	R
Formaldehyde	R	R	Lacquer thinners	C	N
Formic acid	R	N	Lactic acid, 25%	R	R
Freon F11, F12, F113, F114	R	R	Lard oil	R	R
Freon F21, F22	N	N	Lauric acid	R	R
Fruit juices and pulps	R	R	Lauryl chloride	R	R
Fuel oil	C	N	Lauryl sulfate	R	R
Furfural	N	N	Lead salts	R	R
Gallic acid	R	R	Lime sulfur	R	R
Gas, coal, manufactured	N	N	Linoleic acid	R	R
Gas, natural, methane	R	R	Linseed oil	R	R
Gasolines	C	C	Liqueurs	R	R
Gelatin	R	R	Liquors	R	R
Glue, animal	R	R	Lithium salts	R	R
Glycerine (glycerol)	R	R	Lubricating oils	R	R
Glycolic acid	R	R	Machine oil	R	R
Glycols	R	R	Magnesium salts	R	R
Green liquor, paper	R	R	Maleic acid	R	R
Heptane	R	R	Malic acid	R	R
Hexane	R	C	Manganese sulfate	R	R
Hydrazine	N	N	Mercuric salts	R	R
Hydrobromic acid, 20%	R	R	Mercury	R	R
Hydrochloric acid	R	R	Mesityl oxide	N	N
Hydrocyanic acid	R	R	Metallic soaps, aq	R	R
Hydrofluoric acid, 10%	R	C	Methane	R	R
Hydrofluoric acid, 60%	R	C	Methyl acetate	N	N
Hydrofluoric acid, 100%	R	C	Methyl bromide	N	N
Hydrogen	R	R	Methyl cellosolve	N	N
Hydrogen peroxide, 50%	R	R	Methyl chloride	N	N
Hydrogen peroxide, 90%	R	R	Methyl chloroform	N	N
Hydrogen sulfide, aq	R	R	Methyl cyclohexanone	N	N
Hydrogen sulfide, dry	R	R	Methyl methacrylate	R	—
Hydroquinone	R	R	Methyl salicylate	R	R
Hydroxylamine sulfate	R	R	Methyl sulfate	R	C
Hypochlorous acid	R	R	Methyl sulfonic acid	R	R
Iodine, alc	N	N	Methylene bromide	N	N
Iodine, aq, 10%	N	N	Methylene chloride	N	N
Iodine, in KI, 3%, aq	C	N	Methylene iodide	N	N

(continued on next page)

Table A-1 Chemical resistance of PVC pressure water pipe (continued)

Chemical	73°F (23°C)	140°F (60°C)	Chemical	73°F (23°C)	140°F (60°C)
Milk	R	R	Perchloroethylene	C	C
Mineral oil	R	R	Petroleum, refined	R	R
Mixed acids (sulfuric and nitric)	C	N	Petroleum, sour	R	R
Mixed acids (sulfuric and phosphoric)	R	R	Phenol	C	N
Molasses	R	R	Phenylcarbinol	N	N
Monochlorobenzene	N	N	Phenylhydrazine	N	N
Monoethanolamine	N	N	Phenylhydrazine HCl	C	N
Motor oil	R	R	Phosgene, gas	R	C
Naphtha	R	R	Phosgene, liquid	N	N
Naphthalene	N	N	Phosphoric acid	R	R
Nickel salts	R	R	Phosphorus pentoxide	R	C
Nicotine	R	R	Phosphorus, red	R	R
Nicotinic acid	R	R	Phosphorus trichloride	N	N
Nitric acid, 0 to 50%	R	C	Phosphorus, yellow	R	C
Nitric acid, 60%	R	C	Photographic chemicals, aq	R	R
Nitric acid, 70%	R	C	Phthalic acid	C	C
Nitric acid, 80%	C	C	Picric acid	N	N
Nitric acid, 90%	C	N	Plating solutions, metal	R	C
Nitric acid, 100%	N	N	Potassium alkyl xanthates	R	N
Nitric acid, fuming	N	N	Potassium permanganate, 25%	C	C
Nitrobenzene	N	N	Potassium salts, aq	R	R
Nitroglycerine	N	N	Propane	R	R
Nitroglycol	N	N	Propylene dichloride	N	N
Nitropropane	C	C	Propylene glycol	R	R
Nitrous acid	R	C	Propylene oxide	N	N
Nitrous oxide, gas	R	C	Pyridine	N	N
Oils and fats	R	R	Pyrogallic acid	C	C
Oils, vegetable	R	R	Rayon coagulating bath	R	R
Oleic acid	R	R	Salicylaldehyde	C	C
Oleum	N	N	Salicylic acid	R	R
Olive oil	C	—	Sea water	R	R
Oxalic acid	R	R	Selenic acid	R	R
Oxygen, gas	R	R	Sewage, residential	R	R
Ozone, gas	R	C	Silicic acid	R	R
Palmitic acid, 10%	R	R	Silicone oil	R	N
Palmitic acid, 70%	R	N	Silver salts	R	R
Paraffin	R	R	Soaps	R	R
Pentane	C	C	Sodium chlorate	R	C
Peracetic acid, 40%	R	N	Sodium chlorite	R	R
Perchloric acid, 10%	R	C	Sodium dichromate, acid	R	R
Perchloric acid, 70%	R	N	Sodium perborate	R	R

(continued on next page)

Table A-1 Chemical resistance of PVC pressure water pipe (continued)

Chemical	73°F (23°C)	140°F (60°C)	Chemical	73°F (23°C)	140°F (60°C)
Sodium salts, aq	R	R	Titanium tetrachloride	C	N
Stannic chloride	R	R	Toluene	N	N
Stannous chloride	R	R	Tributyl citrate	R	—
Starch	R	R	Tributyl phosphate	N	N
Stearic acid	R	R	Trichloroacetic acid	R	R
Stoddard solvent	N	N	Trichloroethylene	N	N
Sugars, aq	R	R	Tricresyl phosphate	N	N
Sulfite liquor	R	R	Triethanolamine	R	C
Sulfur	R	R	Triethylamine	R	R
Sulfur dioxide, dry	R	R	Trimethyl propane	R	C
Sulfur dioxide, wet	R	C	Turpentine	R	R
Sulfur trioxide, gas, dry	R	R	Urea	R	R
Sulfur trioxide, wet	R	C	Urine	R	R
Sulfuric acid, up to 70%	R	R	Vaseline	N	N
Sulfuric acid, 70 to 90%	R	C	Vegetable oils	R	R
Sulfuric acid, 90 to 100%	C	N	Vinegar	R	R
Sulfurous acid	C	N	Vinyl acetate	N	N
Tall oil	R	R	Water, distilled	R	R
Tannic acid	R	R	Water, fresh	R	R
Tanning liquors	R	R	Water, mine	R	R
Tartaric acid	R	R	Water, salt	R	R
Terpineol	C	C	Water, tap	R	R
Tetrachloroethane	C	C	Whiskey	R	R
Tetraethyl lead	R	C	Wines	R	R
Tetrahydrofuran	N	N	Xylene	N	N
Thionyl chloride	N	N	Zinc salts	R	R
Thread cutting oils	R	—			

Source: PPI TR-19, Plastics Pipe Institute, 1275 K St., N.W., Suite 400, Washington, DC 20005.

Notes: R = generally resistant
C = less resistant than R, but still suitable for some conditions
N = not resistant

This table is provided to aid the designer in decisions regarding exposure to undiluted chemicals except where a diluted concentration is indicated.

Table A-2 General chemical resistance of various gasket materials

Material (ASTM Designation)	General Purpose	—	Non-Oil Resistant	General Purpose	—	Oil Resistant
	Butadiene Styrene (SBR)	Butadiene (BR)	Ethylene Propylene (EPM) Ethylene Propylene Terpolymer (EPDM)	Nitrile (NBR)		Neoprene (CR)
Fluid Resistance Key	SBR	BR	EPM EPDM	NBR		CR
Chemical Group	Poly Butadiene — Butadiene Styrene Copolymer		Ethylene Propylene Copolymer Terpolymer	Butadiene Acrylonitrile Copolymer		Chloroprene Polymer
Generally Resistant to	Most Moderate Chemicals Wet or Dry, Organic Acids, Alcohols, Ketones, Aldehydes		Animal and Vegetable Oils, Ozone, Strong and Oxidizing Chemicals	Many Hydrocarbons, Fats, Oils, Greases, Hydraulic Fluids, Chemicals		Moderate Chemicals and Acids, Ozone, Oils, Fats, Greases, Many Oils and Solvents
Generally Attacked by	Ozone, Strong Acids, Fats, Oils, Greases, Most Hydrocarbons		Mineral Oils and Solvents, Aromatic Hydrocarbons	Ozone (except PVC blends), Ketones, Esters, Aldehydes, Chlorinated and Nitro Hydrocarbons		Strong Oxidizing Acids, Esters, Ketones, Chlorinated, Aromatic and Nitro Hydrocarbons

(continued on next page)

Table A-2 General chemical resistance of various gasket materials (continued)

Fluid Resistance Key	General Purpose —Non-Oil Resistant		General Purpose —Oil Resistant	
	SBR BR	EPM EPDM	NBR	CR
Acetaldehyde	N	R	N	N
Acetamide	N	R	R	C
Acetic acid, 30%	C	R	C	R
Acetic acid, glacial	N	R	N	N
Acetic anhydride	C	C	N	R
Acetone	C	R	N	C
Acetophenone	N	R	N	N
Acetyl chloride	—	—	—	N
Acetylene	C	R	C	C
Acrylonitrile	N	N	N	N
Adipic acid	—	—	R	—
Alkazene	—	N	—	N
Alum-NH$_3$-Cr-K	R	R	R	R
Aluminum acetate	C	R	C	C
Aluminum chloride	N	R	R	R
Aluminum fluoride	R	R	R	R
Aluminum nitrate	R	R	R	R
Aluminum phosphate	R	R	R	R
Aluminum sulfate	C	R	R	R
Ammonia anhydrous	—	R	R	R
Ammonia gas (cold)	R	R	R	R
Ammonia gas (hot)	—	C	—	C
Ammonium carbonate	R	R	N	R
Ammonium chloride	R	R	R	R
Ammonium hydroxide	N	R	N	R
Ammonium nitrate	R	R	R	C
Ammonium nitrite	R	R	R	R
Ammonium persulfate	N	R	N	R
Ammonium phosphate	R	R	R	R
Ammonium sulfate	C	R	R	R
Amyl acetate	N	R	N	N
Amyl alcohol	C	R	C	R
Amyl borate	N	N	R	R
Amyl chloronaphthalene	N	N	—	N
Amyl naphthalene	N	N	N	N
Aniline	N	C	N	N
Aniline dyes	C	C	N	C
Aniline hydrochloride	N	C	C	N
Animal fats	N	C	R	C
Ansul ether	N	N	N	N
Aqua regia	N	N	—	N

(continued on next page)

Table A-2 General chemical resistance of various gasket materials (continued)

	General Purpose —Non-Oil Resistant		General Purpose —Oil Resistant	
Fluid Resistance Key	SBR BR	EPM EPDM	NBR	CR
Arochlor(s)	N	N	N	N
Arsenic acid	R	R	R	R
Arsenic trichloride	—	—	R	R
Askarel	N	N	C	N
Asphalt	N	—	C	N
Barium chloride	R	R	R	R
Barium hydroxide	R	R	R	R
Barium sulfate	R	R	R	R
Barium sulfide	C	R	R	R
Beer	R	R	R	R
Beet sugar liquors	R	R	R	R
Benzaldehyde	N	R	N	N
Benzene	N	N	N	N
Benzenesulfonic acid	—	—	—	R
Benzoic acid	—	—	—	—
Benzyl alcohol	—	C	N	R
Benzyl aenzoate	—	C	—	—
Benzyl chloride	—	—	N	N
Blast furnace gas	N	—	N	N
Bleach solutions	N	R	—	N
Borax C	R	C	R	—
Bordeaux mixture	C	R	—	R
Boric acid	R	R	R	R
Brine	R	R	R	—
Bromine anhydrous	—	—	—	N
Bromine trifluoride	N	N	N	N
Bromine water	—	—	—	C
Bromobenzene	N	N	N	N
Bunker oil	—	—	R	—
Butadiene	N	N	N	C
Butane	N	N	R	R
Butter	N	R	R	C
Butyl acetate	—	C	—	N
Butyl acetyl ricinoleate	—	R	—	C
Butyl acrylate	N	N	—	—
Butyl alcohol	R	C	R	R
Butyl amine	N	N	N	N
Butyl benzoate	—	R	—	N
Butyl carbitol	—	R	R	C
Butyl cellosolve	—	R	N	C
Butyl oleate	N	C	—	N

(continued on next page)

Table A-2 General chemical resistance of various gasket materials (continued)

	General Purpose – Non-Oil Resistant		General Purpose – Oil Resistant	
Fluid Resistance Key	SBR BR	EPM EPDM	NBR	CR
Butyl stearate	N	C	C	—
Butylene	N	N	C	N
Butyraldehyde	N	C	N	N
Calcium acetate	—	R	C	C
Calcium bisulfite	N	N	R	R
Calcium chloride	R	R	R	R
Calcium hydroxide	R	R	R	R
Calcium hypochlorite	N	R	N	N
Calcium nitrate	R	R	R	R
Calcium sulfide	C	R	C	R
Cane sugar liquors	R	R	R	R
Carbamate	N	C	N	C
Carbitol	C	C	C	C
Carbolic acid	N	C	N	N
Carbon bisulfide	—	N	N	N
Carbon dioxide	C	C	R	C
Carbon monoxide	C	R	R	R
Carbon tetrachloride	N	N	N	N
Carbonic acid	C	R	R	R
Castor oil	R	C	R	R
Cellosolve	N	C	—	—
Cellosolve acetate	N	C	N	—
Cellulube	—	R	N	N
Chlorine (Dry)	N	—	—	N
Chlorine (Wet)	N	N	—	N
Chlorine dioxide	—	N	N	N
Chlorine trifluoride	N	N	N	N
1-chloro 1-nitro ethane	N	N	N	N
Chloroacetic acid	—	C	—	—
Chloroacetone	—	R	N	C
Chlorobenzene	N	N	N	N
Chlorobromomethane	N	C	—	N
Chlorobutadiene	N	N	N	N
Chlorododecane	N	N	N	N
Chloroform	N	N	N	N
0-Chloronaphthalene	N	N	N	N
Chlorosulfonic acid	N	N	N	N
Chlorotoluene	N	N	N	N
Chrome plating solutions	N	N	N	N
Chromic acid	N	N	N	N
Citric acid	R	R	R	R

(continued on next page)

118 PVC PIPE—DESIGN AND INSTALLATION

Table A-2 General chemical resistance of various gasket materials (continued)

	General Purpose —Non-Oil Resistant		General Purpose —Oil Resistant	
Fluid Resistance Key	SBR BR	EPM EPDM	NBR	CR
Cobalt chloride	R	R	R	R
Coconut oil	N	R	R	C
Cod liver oil	N	R	R	C
Coke oven gas	N	—	—	—
Copper acetate	—	R	C	C
Copper chloride	R	R	R	R
Copper cyanide	R	R	R	R
Copper sulfate	C	R	R	R
Corn oil	N	N	R	C
Cottonseed oil	N	R	R	C
Creosote	N	N	C	N
Cresol	N	N	N	N
Cresylic acid	N	N	N	N
Cumene	—	—	—	N
Cyclohexane	N	N	R	N
Cyclohexanol	N	N	C	R
Cyclohexanone	—	C	N	N
p-cymene	—	—	—	N
Decalin	N	—	—	N
Decane	N	—	C	N
Denatured alcohol	R	R	R	R
Detergent solutions	C	R	R	R
Developing fluids	C	C	R	R
Diacetone	—	R	—	—
Diacetone alcohol	N	R	N	R
Dibenzyl ether	N	C	N	C
Dibenzyl sebecate	—	C	—	N
Dibutyl amine	N	N	N	N
Dibutyl ether	N	N	N	N
Dibutyl phthalate	N	R	N	N
Dibutyl sebecate	N	C	N	N
o-dichlorobenzene	N	N	N	N
Dichloro-isopropyl ether	N	N	N	N
Didaclohexylamine	N	—	N	—
Diesel oil	N	N	R	C
Diethlamine	C	C	N	N
Diethyl benezene	N	N	N	N
Diethyl ether	N	N	N	N
Diethyl sebecate	—	C	N	N
Diethylene glycol	R	R	R	R
Diisobutylene	—	—	C	N

(continued on next page)

Table A-2 General chemical resistance of various gasket materials (continued)

Fluid Resistance Key	General Purpose —Non-Oil Resistant		General Purpose —Oil Resistant	
	SBR BR	EPM EPDM	NBR	CR
Diisopropyl benzene	N	N	N	N
Diisopropyl ketone	—	R	N	N
Dimethyl aniline	N	C	—	N
Dimethyl formamide	—	—	C	N
Dimethyl phthalate	N	C	N	N
Dinitrotoluene	N	N	N	N
Dioctyl phthalate	—	C	—	N
Dioctyl sebecate	N	C	N	N
Dioxane	—	C	—	—
Dioxolane	N	C	N	—
Dipentene	—	—	C	—
Diphenyl	—	—	—	—
Diphenyl oxides	—	R	—	—
Dowtherm oil	N	N	—	N
Dry cleaning fluids	N	N	N	N
Epichlorohydrin	N	C	—	—
Ethane	N	N	R	C
Ethanolamine	C	C	C	C
Ethyl acetate	N	C	N	N
Ethyl acetoacetate	N	C	N	N
Ethyl acrylate	—	C	—	—
Ethyl alcohol	R	R	R	R
Ethyl benzene	N	N	N	N
Ethyl benzoate	—	C	—	—
Ethyl cellosolve	—	C	—	—
Ethyl cellulose	C	C	—	C
Ethyl chloride	C	R	R	C
Ethyl chlorocarbonate	N	—	—	N
Ethyl chloroformate	—	—	—	N
Ethyl ether	—	N	N	N
Ethyl formate	N	C	NBR	C
Ethyl mercaptan	N	N	N	—
Ethyl oxalate	R	R	N	N
Ethyl pentochlorobenzene	N	N	N	N
Ethyl silicate	C	R	R	R
Ethylene	—	—	R	—
Ethylene chloride	—	N	—	—
Ethylene chlorohydrin	C	—	N	C
Ethylene diamine	C	R	R	R
Ethylene dichloride	N	N	N	N
Ethylene glycol	R	R	R	R

(continued on next page)

Table A-2 General chemical resistance of various gasket materials (continued)

Fluid Resistance Key	General Purpose —Non-Oil Resistant		General Purpose —Oil Resistant	
	SBR BR	EPM EPDM	NBR	CR
Ethylene oxide	—	N	N	N
Ethylene trichloride	—	N	N	N
Fatty acids	N	N	C	C
Ferric chloride	R	R	R	R
Ferric nitrate	R	R	R	R
Ferric sulfate	R	R	R	R
Fish oil	—	—	R	—
Fluorinated cyclic ethers	—	R	—	—
Fluorine (liquid)	—	N	—	—
Fluorobenzene	N	N	N	N
Fluoroboric acid	R	R	R	R
Fluorocarbon oils	—	R	—	—
Fluorolube	N	R	R	R
Fluosilicic acid	—	—	R	R
Formaldehyde	—	R	C	R
Formic acid	R	R	C	R
Freon 11	N	N	R	C
Freon 12	R	C	R	R
Freon 13	R	R	R	R
Freon 13B1	R	R	R	R
Freon 21	N	N	C	—
Freon 22	R	R	N	R
Freon 31	C	R	N	R
Freon 32	R	R	R	R
Freon 112	—	N	C	C
Freon 113	C	N	R	R
Freon 114	R	R	R	R
Freon 114B2	N	N	C	R
Freon 115	R	R	R	R
Freon 142b	R	R	R	R
Freon 152a	R	R	R	R
Freon 218	R	R	R	R
Freon 502	R	—	C	R
Freon BF	N	—	C	C
Freon C316	R	R	R	R
Freon C318	R	R	R	R
Freon MF	C	—	R	N
Freon TA	R	R	R	R
Freon TC	C	C	R	R
Freon TF	C	N	R	R
Freon TMC	N	C	C	C

(continued on next page)

Table A-2 General chemical resistance of various gasket materials (continued)

Fluid Resistance Key	General Purpose —Non-Oil Resistant		General Purpose —Oil Resistant	
	SBR BR	EPM EPDM	NBR	CR
Freon T-P35	R	R	R	R
Freon T-WD602	C	C	C	C
Fuel oil	N	N	R	C
Fufural	N	C	N	C
Fumaric acid	R	—	R	C
Furan, furfuran	N	N	N	N
Gallic acid	C	C	C	C
Gasoline	N	N	R	C
Gelatin	R	R	R	R
Glauber's salt	N	C	—	—
Glucose	R	R	R	R
Glue	R	R	R	R
Glycerin	R	R	R	—
Glycols	R	R	R	R
Green sulfate liquor	C	R	C	C
Halowax oil	N	N	N	N
n-hexaldehyde	N	R	N	R
Hexane	N	N	R	C
n-hexene-1	N	N	C	C
Hexyl alcohol	R	N	R	C
Hydraulic oil (petroleum)	N	N	R	C
Hydrazine	—	R	C	C
Hydrazine (ODMH)	—	—	—	—
Hydrobromic acid	N	R	N	R
Hydrochloric acid (cold) 37%	C	R	C	C
Hydrochloric acid (hot) 37%	N	N	N	N
Hydrocyanic acid	C	R	C	C
Hydrofluoric acid-Anhydrous	N	C	—	—
Hydrofluoric acid (conc.) cold	N	C	N	C
Hydrofluoric acid (conc.) hot	N	N	N	N
Hydrofluosilicic acid	C	R	C	N
Hydrogen gas	C	R	R	R
Hydrogen peroxide (90%)	N	N	N	—
Hydrogen sulfide (wet, cold)	N	R	N	R
Hydrogen sulfide (wet, hot)	N	R	N	C
Hydroquinone	C	—	N	—
Hypochlorous acid	C	C	N	—
Iodine pentafluoride	N	N	N	N
Iodoform	R	—	—	—
Isobutyl alcohol	C	R	C	R
Isooctane	N	N	R	C

(continued on next page)

Table A-2 General chemical resistance of various gasket materials (continued)

	General Purpose —Non-Oil Resistant		General Purpose —Oil Resistant	
Fluid Resistance Key	SBR BR	EPM EPDM	NBR	CR
Isophorone	—	R	N	—
Isopropyl acetate	—	R	N	N
Isopropyl alcohol	C	R	C	R
Isopropyl chloride	N	N	N	—
Isopropyl ether	N	N	C	C
Kerosene	N	N	R	N
Lacquer solvents	N	N	N	N
Lacquers	N	N	N	N
Lactic acid	R	R	R	R
Lard	N	N	R	N
Lavender oil	N	N	C	N
Lead acetate	—	R	C	C
Lead nitrate	R	R	R	R
Lead sulfamate	C	R	C	R
Lime bleach	R	R	R	C
Lime sulfur	N	R	N	R
Lindol	—	R	—	N
Linoleic acid	—	N	C	N
Linseed oil	N	C	R	C
Liquefied petroleum gas	N	N	R	C
Lubricating oils (petroleum)	N	N	R	C
Lye	C	R	C	C
Magnesium chloride	R	R	R	R
Magnesium hydroxide	C	R	C	R
Magnesium sulfate	C	R	R	R
Maleic acid	C	N	—	—
Maleic anhydride	C	N	—	—
Malic acid	C	N	R	C
Mercuric chloride	R	R	R	R
Mercury	R	R	R	R
Mesityl oxide	N	C	N	N
Methane	N	N	R	C
Methyl acetate	N	C	N	C
Methyl acrylate	N	C	N	C
Methyl alcohol	R	R	R	R
Methyl bromide	—	—	C	N
Methyl butyl ketone	N	R	N	N
Methyl cellosolve	N	C	—	C
Methyl chloride	N	N	N	N
Methyl cyclopentane	N	N	—	N
Methyl ethyl ketone	N	R	N	N

(continued on next page)

Table A-2 General chemical resistance of various gasket materials (continued)

Fluid Resistance Key	General Purpose —Non-Oil Resistant		General Purpose —Oil Resistant	
	SBR BR	EPM EPDM	NBR	CR
Methyl formate	N	C	N	C
Methyl isobutyl ketone	N	N	N	N
Methyl methacrylate	N	N	N	N
Methyl oleate	N	C	N	N
Methyl salicylate	—	C	—	N
Methylacrylic acid	N	C	—	C
Methylene chloride	N	N	N	N
Milk	R	R	R	R
Mineral oil	N	N	R	C
Monochlorobenzene	N	N	N	N
Monoethanolamine	C	C	N	N
Monomethyl aniline	N	—	N	N
Monomethylether	C	R	R	R
Monovinyl acetylene	C	R	R	C
Mustard gas	—	R	—	R
Naphtha	N	N	N	N
Naphthalene	N	N	N	N
Naphthenic acid	N	N	C	—
Natural gas	N	N	R	R
Neatsfoot oil	N	C	R	—
Neville acid	N	C	N	N
Nickel acetate	—	R	C	C
Nickel chloride	R	R	R	R
Nickel sulfate	C	R	R	R
Niter cake	R	R	R	R
Nitric acid-conc.	N	N	N	N
Nitric acid-dilute	N	C	N	R
Nitric acid-red fuming	N	N	N	N
Nitrobenzene	N	N	N	N
Nitrobenzine	—	N	—	N
Nitroethane	C	C	N	N
Nitrogen	R	R	R	R
Nitrogen tetroxide	N	N	N	N
Nitromethane	C	C	N	N
Octachlorotoluene	N	N	N	N
Octadecane	N	N	R	C
n-octane	N	N	—	—
Octyl alcohol	C	R	C	R
o-dichlorobenzene	—	—	N	N
Oleic acid	N	C	N	N
Oleum spirits	—	—	C	N

(continued on next page)

Table A-2 General chemical resistance of various gasket materials (continued)

	General Purpose —Non-Oil Resistant		General Purpose —Oil Resistant	
Fluid Resistance Key	SBR BR	EPM EPDM	NBR	CR
Olive oil	N	C	R	C
Oxalic acid	C	R	C	C
Oxygen-200-400°F	N	N	N	N
Oxygen-cold	C	R	C	C
Ozone	N	R	N	C
Paint thinner, duco	N	N	—	—
Palmitic acid	C	C	R	C
Peanut oil	N	N	R	C
Perchloric acid	—	C	—	R
Perchloroethylene	N	N	N	N
Petroleum-above 250	N	N	N	N
Petroleum-below 250	N	N	R	C
Phenol	—	C	—	N
Phenyl ethyl ether	N	N	N	N
Phenyl hydrazine	C	N	N	N
Phenylbenzene	N	N	N	N
Phorone	—	C	—	—
Phosphoric acid-20%	N	R	C	C
Phosphoric acid-45%	N	C	N	C
Phosphorous trichloride	N	R	N	N
Pickling solution	—	N	—	—
Picric acid	C	C	C	R
Pine oil	N	N	C	N
Pinene	N	N	C	C
Piperidine	N	N	N	N
Plating solution-chrome	N	R	—	—
Plating solution-others	—	R	R	—
Polyvinyl acetate emulsion	—	R	—	C
Potassium acetate	—	R	C	C
Potassium chloride	R	R	R	R
Potassium cupro cyanide	R	R	R	R
Potassium cyanide	R	R	R	R
Potassium dichromate	C	R	R	R
Potassium hydroxide	C	R	C	R
Potassium nitrate	R	R	R	R
Potassium sulfate	C	R	R	R
Producer gas	N	N	R	C
Propane	N	N	R	R
Propyl acetate	N	C	N	N
n-propyl acetate	N	R	N	—
Propyl alcohol	R	R	R	R

(continued on next page)

Table A-2 General chemical resistance of various gasket materials (continued)

Fluid Resistance Key	General Purpose —Non-Oil Resistant		General Purpose —Oil Resistant	
	SBR BR	EPM EPDM	NBR	CR
Propyl nitrate	—	C	—	—
Propylene	N	N	N	N
Propylene oxide	—	C	—	N
Pydrauls	N	C	N	N
Pyranol	N	N	R	N
Pyridine	N	C	N	N
Pyroligneous acid	—	C	—	C
Pyrrole	N	N	N	N
Radiation	C	C	C	C
Rapeseed oil	N	R	C	C
Red oil	N	N	R	C
Sal ammoniac	R	R	R	R
Salicylic acid	C	R	R	—
Salt water	R	R	R	R
Silicate esters	N	N	C	R
Silicone greases	R	R	R	R
Silicone oils	R	R	R	R
Silver nitrate	R	R	C	R
Skydrol 500	N	R	N	N
Skydrol 7000	N	R	N	N
Soap solutions	C	R	R	R
Soda ash	R	R	R	R
Sodium acetate	N	R	C	C
Sodium bicarbonate	R	R	R	R
Sodium bisulfite	C	R	R	R
Sodium borate	R	R	R	R
Sodium chloride	R	R	R	R
Sodium cyanide	R	R	R	R
Sodium hydroxide	R	R	C	R
Sodium hypochlorite	N	C	C	C
Sodium metaphosphate	R	R	R	C
Sodium nitrate	C	R	C	R
Sodium perborate	C	R	C	C
Sodium peroxide	C	R	C	C
Sodium phosphate	R	R	R	R
Sodium silicate	R	R	R	R
Sodium sulfate	C	R	R	R
Sodium thiosulfate	C	R	C	R
Soybean oil	N	N	R	C
Stannic(ous) chloride	R	C	R	R
Steam over 300°F	N	C	N	N

(continued on next page)

Table A-2 General chemical resistance of various gasket materials (continued)

	General Purpose—Non-Oil Resistant		General Purpose—Oil Resistant	
Fluid Resistance Key	SBR BR	EPM EPDM	NBR	CR
Steam under 300°F	N	R	N	N
Stearic acid	C	C	C	C
Stoddard solvent	N	N	R	N
Styrene	N	N	N	N
Sucrose solution	R	R	R	R
Sulfite liquors	C	C	C	C
Sulfur	N	R	N	R
Sulfur chloride	N	N	N	N
Sulfur dioxide	N	R	N	N
Sulfur hexafluoride	R	R	R	R
Sulfur trioxide	N	C	N	N
Sulfuric acid (20% oleum)	N	N	N	N
Sulfuric acid (conc.)	N	C	N	N
Sulfuric acid (dilute)	N	C	N	C
Sulfurous acid	C	C	C	C
Tannic acid	C	R	R	R
Tar, bituminous	N	N	C	N
Tartaric acid	C	C	R	C
Terpincol	N	N	C	N
Tertiary butyl alcohol	C	C	C	C
Tertiary butyl catechol	N	C	N	C
Tertiary butyl mercaptan	N	N	N	N
Tetrabromomethane	N	N	N	—
Tetrabutyl titanate	C	R	C	R
Tetrachloroethylene	N	N	N	—
Tetraethyl lead	N	N	C	N
Tetrahydrofuran	N	C	—	—
Tetralin	N	N	N	N
Thionyl chloride	N	N	—	N
Titanium tetrachloride	N	N	N	N
Toluene	N	N	N	N
Toluene diisocyanate	N	R	—	N
Transformer oil	N	N	R	C
Transmission fluid type A	N	N	R	C
Tributoxy ethyl phosphate	C	R	N	N
Tributyl mercaptan	N	N	N	N
Tributyl phosphate	N	R	N	N
Trichloroacetic acid	C	C	C	C
Trichloroethane	N	N	N	N
Trichloroethylene	N	N	N	N
Tricresyl phosphate	N	R	N	N

(continued on next page)

Table A-2 General chemical resistance of various gasket materials (continued)

Fluid Resistance Key	General Purpose —Non-Oil Resistant		General Purpose —Oil Resistant	
	SBR BR	EPM EPDM	NBR	CR
Triethanol amine	C	C	N	R
Triethyl aluminum	—	—	—	—
Triethyl borane	—	—	—	—
Trinitrotoluene	N	N	N	C
Triocetin	N	R	C	C
Trioctyl phosphate	N	R	N	N
Trioryl phosphate	N	R	N	N
Tung oil	N	N	R	C
Turbine oil	N	N	C	C
Turpentine	N	N	R	N
Unsymmetrical dimethyl	—	R	C	C
Varnish	N	N	C	N
Vegetable oils	N	R	R	C
Versilube	R	R	R	R
Vinegar	C	R	C	R
Vinyl chloride	—	C	—	N
Wagner 21B fluid	R	R	N	R
Water	R	R	R	R
Whiskey, wines	R	R	R	R
White oil	N	N	R	C
White pine oil	N	N	C	N
Wood oil	N	N	R	C
Xylene	N	N	N	N
Xylidenes	N	N	N	N
Zeolites	R	R	R	R
Zinc acetate	N	R	C	C
Zinc chloride	R	R	R	R
Zinc sulfate	C	R	R	R

Source: The Los Angeles Rubber Group, Inc. Adapted from 1970 Yearbook and Directory.

Notes: N = not resistant
R = generally resistant
C = less resistant than R, but still suitable for some conditions

This table is provided to aid the designer in decisions as to transporting/conveyance of undiluted chemicals. A indicates insufficient test data to provide a rating.

Chemical resistance data are provided as a guide only. Information is based primarily on immersion of unstressed strips in chemicals and to a lesser degree on field experience.

This page intentionally blank.

Appendix **B**

Flow Friction Loss Tables

Table B-1 Flow friction loss, AWWA C900 PVC pipe

Flow (gpm)	4 in. CIOD (AWWA C900) DR 25 Actual OD 4.800 in., Pressure Class 100			4 in. CIOD (AWWA C900) DR 18 Actual OD 4.800 in., Pressure Class 150			4 in. CIOD (AWWA C900) DR 14 Actual OD 4.800 in., Pressure Class 200		
	Velocity (ft/sec)	Loss of Head (ft/100 ft)	Pressure Drop (psi/100 ft)	Velocity (ft/sec)	Loss of Head (ft/100 ft)	Pressure Drop (psi/100 ft)	Velocity (ft/sec)	Loss of Head (ft/100 ft)	Pressure Drop (psi/100 ft)
20	0.423	0.019	0.008	0.456	0.023	0.010	0.493	0.027	0.012
25	0.529	0.028	0.012	0.570	0.034	0.015	0.616	0.041	0.018
30	0.635	0.040	0.017	0.684	0.048	0.021	0.739	0.058	0.025
35	0.741	0.053	0.023	0.798	0.063	0.027	0.862	0.077	0.033
40	0.847	0.068	0.029	0.912	0.081	0.035	0.985	0.098	0.043
45	0.953	0.085	0.037	1.025	0.101	0.044	1.108	0.122	0.053
50	1.058	0.103	0.044	1.139	0.123	0.053	1.231	0.148	0.064
60	1.270	0.144	0.062	1.367	0.172	0.075	1.478	0.208	0.090
70	1.482	0.192	0.083	1.595	0.229	0.099	1.724	0.277	0.120
75	1.588	0.218	0.094	1.709	0.260	0.113	1.847	0.315	0.136
80	1.694	0.245	0.106	1.823	0.294	0.127	1.970	0.354	0.153
90	1.905	0.305	0.132	2.051	0.365	0.158	2.216	0.441	0.191
100	2.117	0.371	0.161	2.279	0.444	0.192	2.463	0.536	0.232
125	2.646	0.561	0.243	2.849	0.671	0.290	3.078	0.810	0.351
150	3.175	0.786	0.340	3.418	0.940	0.407	3.694	1.135	0.492
175	3.705	1.045	0.453	3.988	1.251	0.542	4.310	1.511	0.654
200	4.234	1.339	0.580	4.558	1.602	0.693	4.925	1.934	0.837
250	5.292	2.024	0.876	5.697	2.422	1.048	6.156	2.924	1.266
300	6.351	2.837	1.228	6.837	3.394	1.469	7.388	4.099	1.774
350	7.409	3.774	1.634	7.976	4.516	1.955	8.619	5.453	2.361
400	8.468	4.833	2.092	9.115	5.783	2.503	9.850	6.983	3.023
450	9.526	6.011	2.602	10.255	7.192	3.113	11.082	8.685	3.760
500	10.584	7.306	3.163	11.394	8.742	3.784	12.313	10.557	4.570
600	12.701	10.241	4.433	13.673	12.253	5.304	14.775	14.797	6.405
700	14.818	13.625	5.898	15.952	16.302	7.057	17.238	19.686	8.522

(continued on next page)

APPENDIX B   131

Table B-1   Flow friction loss, AWWA C900 PVC pipe (continued)

Flow (gpm)	6 in. CIOD (AWWA C900) DR 25 Actual OD 6.900 in., Pressure Class 100			6 in. CIOD (AWWA C900) DR 18 Actual OD 6.900 in., Pressure Class 150			6 in. CIOD (AWWA C900) DR 14 Actual OD 6.900 in., Pressure Class 200		
	Velocity (ft/sec)	Loss of Head (ft/100 ft)	Pressure Drop (psi/100 ft)	Velocity (ft/sec)	Loss of Head (ft/100 ft)	Pressure Drop (psi/100 ft)	Velocity (ft/sec)	Loss of Head (ft/100 ft)	Pressure Drop (psi/100 ft)
50	0.512	0.018	0.008	0.551	0.021	0.009	0.596	0.025	0.011
60	0.615	0.025	0.011	0.661	0.029	0.013	0.715	0.036	0.015
70	0.717	0.033	0.014	0.772	0.039	0.017	0.834	0.047	0.020
75	0.768	0.037	0.016	0.827	0.044	0.019	0.894	0.054	0.023
80	0.820	0.042	0.018	0.882	0.050	0.022	0.953	0.061	0.026
90	0.922	0.052	0.023	0.992	0.062	0.027	1.073	0.075	0.033
100	1.024	0.063	0.027	1.102	0.076	0.033	1.192	0.092	0.040
125	1.281	0.096	0.042	1.378	0.115	0.050	1.490	0.139	0.060
150	1.537	0.134	0.058	1.653	0.161	0.070	1.788	0.194	0.084
175	1.793	0.179	0.077	1.929	0.214	0.093	2.085	0.258	0.112
200	2.049	0.229	0.099	2.204	0.274	0.118	2.383	0.331	0.143
250	2.561	0.346	0.150	2.756	0.414	0.179	2.979	0.500	0.217
300	3.073	0.485	0.210	3.307	0.580	0.251	3.575	0.701	0.304
350	3.585	0.646	0.279	3.858	0.771	0.334	4.717	0.933	0.404
400	4.098	0.827	0.358	4.409	0.988	0.428	4.767	1.194	0.517
450	4.610	1.028	0.445	4.960	1.229	0.532	5.363	1.486	0.643
500	5.122	1.250	0.541	5.511	1.493	0.647	5.958	1.806	0.782
600	6.146	1.752	0.758	6.613	2.093	0.906	7.150	2.531	1.096
700	7.171	2.331	1.009	7.716	2.785	1.206	8.342	3.367	1.458
800	8.195	2.985	1.292	8.818	3.566	1.544	9.534	4.312	1.867
1,000	10.244	4.512	1.953	11.022	5.391	2.334	11.917	6.519	2.822

(continued on next page)

Table B-1 Flow friction loss, AWWA C900 PVC pipe (continued)

Flow (gpm)	8 in. CIOD (AWWA C900) DR 25 Actual OD 9.050 in., Pressure Class 100			8 in. CIOD (AWWA C900) DR 18 Actual OD 9.050 in., Pressure Class 150			8 in. CIOD (AWWA C900) DR 14 Actual OD 9.050 in., Pressure Class 200		
	Velocity (ft/sec)	Loss of Head (ft/100 ft)	Pressure Drop (psi/100 ft)	Velocity (ft/sec)	Loss of Head (ft/100 ft)	Pressure Drop (psi/100 ft)	Velocity (ft/sec)	Loss of Head (ft/100 ft)	Pressure Drop (psi/100 ft)
100	0.595	0.017	0.007	0.641	0.020	0.009	0.692	0.024	0.011
125	0.744	0.026	0.011	0.801	0.031	0.013	0.866	0.037	0.016
150	0.893	0.036	0.016	0.961	0.043	0.019	1.039	0.052	0.022
175	1.042	0.048	0.021	1.122	0.057	0.025	1.212	0.069	0.030
200	1.191	0.061	0.026	1.282	0.073	0.032	1.385	0.088	0.038
250	1.489	0.092	0.040	1.602	0.111	0.048	1.731	0.134	0.058
300	1.786	0.130	0.056	1.923	0.155	0.067	2.077	0.187	0.081
350	2.084	0.172	0.075	2.243	0.206	0.089	2.424	0.249	0.108
400	2.382	0.221	0.096	2.564	0.264	0.114	2.770	0.319	0.138
450	2.680	0.275	0.119	2.884	0.329	0.142	3.116	0.397	0.172
500	2.977	0.334	0.145	3.204	0.399	0.173	3.462	0.482	0.209
600	3.573	0.468	0.203	3.845	0.560	0.242	4.155	0.676	0.292
700	4.168	0.623	0.270	4.486	0.745	0.322	4.847	0.899	0.389
800	4.764	0.797	0.345	5.127	0.954	0.413	5.540	1.151	0.498
1,000	5.954	1.205	0.522	6.409	1.441	0.624	6.924	1.740	0.753
1,200	7.145	1.690	0.731	7.691	2.020	0.875	8.309	2.439	1.056
1,400	8.336	2.248	0.973	8.972	2.688	1.164	9.694	3.245	1.405
1,600	9.527	2.878	1.246	10.254	3.442	1.490	11.079	4.155	1.799
2,000	11.909	4.351	1.884	12.818	5.204	2.253	13.849	6.282	2.719

(continued on next page)

Table B-1 Flow friction loss, AWWA C900 PVC pipe (continued)

Flow (gpm)	10 in. CIOD (AWWA C900) DR 25 Actual OD 11.100 in., Pressure Class 100			10 in. CIOD (AWWA C900) DR 18 Actual OD 11.100 in., Pressure Class 150			10 in. CIOD (AWWA C900) DR 14 Actual OD 11.100 in., Pressure Class 200		
	Velocity (ft/sec)	Loss of Head (ft/100 ft)	Pressure Drop (psi/100 ft)	Velocity (ft/sec)	Loss of Head (ft/100 ft)	Pressure Drop (psi/100 ft)	Velocity (ft/sec)	Loss of Head (ft/100 ft)	Pressure Drop (psi/100 ft)
175	0.693	0.018	0.008	0.746	0.021	0.009	0.806	0.026	0.011
200	0.792	0.023	0.010	0.852	0.027	0.012	0.921	0.033	0.014
250	0.990	0.034	0.015	1.065	0.041	0.018	1.151	0.049	0.021
300	1.188	0.048	0.021	1.278	0.057	0.025	1.381	0.069	0.030
350	1.385	0.064	0.028	1.491	0.076	0.033	1.612	0.092	0.040
400	1.583	0.082	0.035	1.704	0.098	0.042	1.842	0.118	0.051
450	1.781	0.102	0.044	1.917	0.122	0.053	2.072	0.147	0.064
500	1.979	0.124	0.054	2.130	0.148	0.064	2.302	0.179	0.077
600	2.375	0.173	0.075	2.556	0.207	0.090	2.763	0.250	0.108
700	2.771	0.231	0.100	2.982	0.276	0.119	3.223	0.333	0.144
800	3.167	0.295	0.128	3.409	0.353	0.153	3.684	0.427	0.185
1,000	3.958	0.446	0.193	4.261	0.534	0.231	4.605	0.645	0.279
1,200	4.750	0.626	0.271	5.113	0.748	0.324	5.526	0.904	0.391
1,400	5.542	0.832	0.360	5.965	0.996	0.431	6.447	1.203	0.521
1,600	6.333	1.066	0.461	6.817	1.275	0.552	7.368	1.540	0.667
2,000	7.917	1.612	0.698	8.521	1.927	0.834	9.210	2.328	1.008
2,500	9.896	2.436	1.055	10.652	2.914	1.261	11.512	3.520	1.524
3,000	11.875	3.415	1.478	12.782	4.084	1.768	13.814	4.934	2.136

(continued on next page)

Table B-1   Flow friction loss, AWWA C900 PVC pipe (continued)

Flow (gpm)	12 in. CIOD (AWWA C900) DR 25 Actual OD 13.200 in., Pressure Class 100			12 in. CIOD (AWWA C900) DR 18 Actual OD 13.200 in., Pressure Class 150			12 in. CIOD (AWWA C900) DR 14 Actual OD 13.200 in., Pressure Class 200		
	Velocity (ft/sec)	Loss of Head (ft/100 ft)	Pressure Drop (psi/100 ft)	Velocity (ft/sec)	Loss of Head (ft/100 ft)	Pressure Drop (psi/100 ft)	Velocity (ft/sec)	Loss of Head (ft/100 ft)	Pressure Drop (psi/100 ft)
300	0.840	0.021	0.009	0.904	0.025	0.011	0.977	0.030	0.013
350	0.980	0.027	0.012	1.054	0.033	0.014	1.140	0.040	0.017
400	1.120	0.035	0.015	1.205	0.042	0.018	1.302	0.051	0.022
450	1.260	0.044	0.019	1.355	0.052	0.023	1.465	0.063	0.027
500	1.400	0.053	0.023	1.506	0.064	0.028	1.628	0.077	0.033
600	1.679	0.075	0.032	1.807	0.089	0.039	1.954	0.108	0.047
700	1.959	0.099	0.043	2.108	0.119	0.051	2.279	0.143	0.062
800	2.239	0.127	0.055	2.410	0.152	0.066	2.605	0.184	0.079
1,000	2.799	0.192	0.083	3.012	0.230	0.099	3.256	0.278	0.120
1,200	3.359	0.269	0.117	3.614	0.322	0.139	3.907	0.389	0.168
1,400	3.919	0.358	0.155	4.217	0.428	0.185	4.559	0.518	0.224
1,600	4.479	0.459	0.199	4.819	0.548	0.237	5.210	0.663	0.287
2,000	5.598	0.694	0.300	6.024	0.829	0.359	6.512	1.002	0.434
2,500	6.998	1.049	0.454	7.530	1.253	0.543	8.140	1.515	0.656
3,000	8.397	1.470	0.636	9.036	1.757	0.761	9.768	2.123	0.919
3,500	9.797	1.955	0.846	10.542	2.337	1.012	11.397	2.825	1.223
4,000	11.196	2.504	1.084	12.048	2.993	1.296	13.025	3.618	1.566
4,500	12.596	3.114	1.348	13.554	3.722	1.611	14.653	4.499	1.948

Table B-2 Flow friction loss, AWWA C905 pipe

Flow (gpm)	14 in. CIOD (AWWA C905) DR 41 Actual OD 15.30 in. Pressure Rating 100 psi			14 in. CIOD (AWWA C905) DR 32.5 Actual OD 15.30 in. Pressure Rating 125 psi			14 in. CIOD (AWWA C905) DR 25 Actual OD 15.30 in. Pressure Rating 165 psi			14 in. CIOD (AWWA C905) DR 21 Actual OD 15.30 in. Pressure Rating 200 psi			14 in. CIOD (AWWA C905) DR 18 Actual OD 15.30 in. Pressure Rating 235 psi			14 in. CIOD (AWWA C905) DR 14 Actual OD 15.30 in. Pressure Rating 305 psi		
	Velocity (ft/sec)	Loss of Head (ft/100 ft)	Pressure Drop (psi/100 ft)	Velocity (ft/sec)	Loss of Head (ft/100 ft)	Pressure Drop (psi/100 ft)	Velocity (ft/sec)	Loss of Head (ft/100 ft)	Pressure Drop (psi/100 ft)	Velocity (ft/sec)	Loss of Head (ft/100 ft)	Pressure Drop (psi/100 ft)	Velocity (ft/sec)	Loss of Head (ft/100 ft)	Pressure Drop (psi/100 ft)	Velocity (ft/sec)	Loss of Head (ft/100 ft)	Pressure Drop (psi/100 ft)
700	1.360	0.041	0.018	1.398	0.044	0.019	1.458	0.048	0.021	1.512	0.053	0.023	1.569	0.058	0.025	1.696	0.070	0.030
900	1.748	0.065	0.028	1.797	0.070	0.030	1.875	0.077	0.033	1.943	0.084	0.036	2.018	0.092	0.040	2.181	0.111	0.048
1,100	2.137	0.094	0.041	2.197	0.101	0.044	2.292	0.112	0.048	2.375	0.122	0.053	2.466	0.134	0.058	2.666	0.162	0.070
1,300	2.525	0.128	0.056	2.596	0.137	0.059	2.708	0.152	0.066	2.807	0.166	0.072	2.915	0.182	0.079	3.151	0.220	0.095
1,500	2.914	0.167	0.072	2.996	0.179	0.078	3.125	0.198	0.086	3.239	0.217	0.094	3.363	0.237	0.103	3.635	0.287	0.124
1,700	3.302	0.211	0.091	3.395	0.226	0.098	3.542	0.250	0.108	3.671	0.273	0.119	3.812	0.299	0.130	4.120	0.362	0.157
1,900	3.691	0.259	0.112	3.794	0.277	0.120	3.958	0.308	0.133	4.103	0.336	0.146	4.260	0.368	0.159	4.605	0.445	0.193
2,100	4.079	0.312	0.135	4.194	0.334	0.145	4.375	0.370	0.160	4.535	0.404	0.175	4.708	0.443	0.192	5.089	0.535	0.232
2,300	4.468	0.369	0.160	4.593	0.395	0.171	4.792	0.438	0.190	4.967	0.479	0.207	5.157	0.524	0.227	5.574	0.634	0.275
2,500	4.856	0.431	0.187	4.993	0.461	0.200	5.208	0.511	0.221	5.398	0.558	0.242	5.605	0.611	0.265	6.059	0.739	0.321
2,700	5.244	0.497	0.215	5.392	0.532	0.230	5.625	0.590	0.255	5.830	0.644	0.279	6.054	0.705	0.305	6.543	0.853	0.370
2,900	5.633	0.568	0.246	5.792	0.607	0.263	6.042	0.637	0.291	6.262	0.735	0.319	6.502	0.805	0.348	7.028	0.973	0.422
3,000	5.827	0.604	0.262	5.991	0.647	0.280	6.250	0.717	0.310	6.478	0.783	0.339	6.726	0.857	0.371	7.270	1.036	0.449
3,500	6.798	0.804	0.348	6.990	0.860	0.372	7.292	0.953	0.413	7.558	1.041	0.451	7.847	1.140	0.493	8.482	1.379	0.598
4,000	7.770	1.030	0.446	7.988	1.102	0.477	8.333	1.221	0.528	8.638	1.333	0.578	8.968	1.460	0.632	9.694	1.766	0.765
4,500	8.741	1.281	0.554	8.897	1.370	0.593	9.375	1.518	0.657	9.717	1.659	0.719	10.089	1.815	0.786	10.906	2.196	0.952
5,000	9.712	1.556	0.674	9.985	1.665	0.721	10.416	1.846	0.799	10.797	2.016	0.874	11.210	2.207	0.955	12.117	2.669	1.157
5,500	10.683	1.857	0.804	10.984	1.987	0.860	11.458	2.202	0.953	11.877	2.405	1.043	12.332	2.633	1.140	13.329	3.184	1.380

(continued on next page)

136 PVC PIPE—DESIGN AND INSTALLATION

Table B-2 Flow friction loss, AWWA C905 pipe (continued)

Flow (gpm)	16 in. CIOD (AWWA C905) DR 41 Actual OD 17.40 in. Pressure Rating 100 psi			16 in. CIOD (AWWA C905) DR 32.5 Actual OD 17.40 in. Pressure Rating 125 psi			16 in. CIOD (AWWA C905) DR 25 Actual OD 17.40 in. Pressure Rating 165 psi			16 in. CIOD (AWWA C905) DR 21 Actual OD 17.40 in. Pressure Rating 200 psi			16 in. CIOD (AWWA C905) DR 18 Actual OD 17.40 in. Pressure Rating 235 psi			16 in. CIOD (AWWA C905) DR 14 Actual OD 17.40 in. Pressure Rating 305 psi		
	Velocity (ft/sec)	Loss of Head (ft/100 ft)	Pressure Drop (psi/100 ft)	Velocity (ft/sec)	Loss of Head (ft/100 ft)	Pressure Drop (psi/100 ft)	Velocity (ft/sec)	Loss of Head (ft/100 ft)	Pressure Drop (psi/100 ft)	Velocity (ft/sec)	Loss of Head (ft/100 ft)	Pressure Drop (psi/100 ft)	Velocity (ft/sec)	Loss of Head (ft/100 ft)	Pressure Drop (psi/100 ft)	Velocity (ft/sec)	Loss of Head (ft/100 ft)	Pressure Drop (psi/100 ft)
900	1.352	0.035	0.015	1.390	0.037	0.016	1.450	0.041	0.018	1.503	0.045	0.020	1.560	0.049	0.021	1.686	0.060	0.026
1,200	1.802	0.059	0.026	1.853	0.063	0.027	1.933	0.070	0.030	2.003	0.077	0.033	2.080	0.084	0.036	2.249	0.102	0.044
1,500	2.253	0.090	0.039	2.316	0.096	0.041	2.416	0.106	0.046	2.504	0.116	0.050	2.601	0.127	0.055	2.811	0.154	0.067
1,800	2.703	0.125	0.054	2.779	0.134	0.058	2.900	0.149	0.064	3.005	0.163	0.070	3.121	0.178	0.077	3.373	0.215	0.093
2,100	3.154	0.167	0.072	3.242	0.179	0.077	3.383	0.198	0.086	3.506	0.216	0.094	3.641	0.237	0.102	3.935	0.286	0.124
2,400	3.604	0.214	0.093	3.705	0.229	0.099	3.866	0.254	0.110	4.007	0.277	0.120	4.161	0.303	0.131	4.497	0.367	0.159
2,700	4.055	0.266	0.115	4.169	0.284	0.123	4.350	0.315	0.137	4.508	0.344	0.149	4.681	0.377	0.163	5.059	0.456	0.198
3,000	4.505	0.323	0.140	4.632	0.346	0.150	4.833	0.383	0.166	5.009	0.419	0.181	5.201	0.458	0.198	5.621	0.554	0.240
3,300	4.956	0.386	0.167	5.095	0.412	0.179	5.316	0.457	0.198	5.510	0.499	0.216	5.721	0.547	0.237	6.184	0.661	0.287
3,500	5.256	0.430	0.186	5.404	0.460	0.199	5.638	0.510	0.221	5.844	0.557	0.241	6.068	0.610	0.264	6.558	0.737	0.320
4,000	6.007	0.551	0.238	6.176	0.589	0.255	6.444	0.653	0.283	6.678	0.713	0.309	6.935	0.781	0.338	7.495	0.944	0.409
4,500	6.758	0.685	0.296	6.948	0.733	0.317	7.249	0.812	0.352	7.513	0.887	0.385	7.802	0.971	0.420	8.432	1.174	0.509
5,000	7.509	0.832	0.360	7.720	0.890	0.385	8.055	0.987	0.427	8.348	1.078	0.467	8.669	1.180	0.511	9.369	1.428	0.619
5,500	8.260	0.993	0.430	8.492	1.062	0.460	8.860	1.178	0.510	9.183	1.286	0.558	9.535	1.408	0.610	10.306	1.703	0.738
6,000	9.010	1.167	0.505	9.264	1.248	0.540	9.666	1.384	0.599	10.017	1.511	0.655	10.402	1.655	0.716	11.243	2.001	0.867
6,500	9.761	1.353	0.586	10.036	1.447	0.627	10.471	1.605	0.695	10.852	1.753	0.760	11.269	1.919	0.831	12.180	2.321	1.006
7,000	10.512	1.552	0.672	10.818	1.660	0.719	11.277	1.841	0.797	11.687	2.010	0.872	12.136	2.201	0.953	13.117	2.662	1.154
7,500	11.263	1.764	0.763	11.580	1.887	0.817	12.082	2.092	0.906	12.522	2.284	0.990	13.003	2.501	1.083	14.053	3.025	1.311

(continued on next page)

Table B-2 Flow friction loss, AWWA C905 pipe (continued)

Flow (gpm)	18 in. CIOD (AWWA C905) DR 51 Actual OD 19.50 in. Pressure Rating 80 psi		18 in. CIOD (AWWA C905) DR 41 Actual OD 19.50 in. Pressure Rating 100 psi			18 in. CIOD (AWWA C905) DR 32.5 Actual OD 19.50 in. Pressure Rating 125 psi			18 in. CIOD (AWWA C905) DR 25 Actual OD 19.50 in. Pressure Rating 165 psi			18 in. CIOD (AWWA C905) DR 21 Actual OD 19.50 in. Pressure Rating 200 psi			18 in. CIOD (AWWA C905) DR 18 Actual OD 19.50 in. Pressure Rating 235 psi			18 in. CIOD (AWWA C905) DR 14 Actual OD 19.50 in. Pressure Rating 305 psi		
	Velocity (ft/sec)	Loss of Head (ft/100 ft)	Pressure Drop (psi/100 ft)	Velocity (ft/sec)	Loss of Head (ft/100 ft)	Pressure Drop (psi/100 ft)	Velocity (ft/sec)	Loss of Head (ft/100 ft)	Pressure Drop (psi/100 ft)	Velocity (ft/sec)	Loss of Head (ft/100 ft)	Pressure Drop (psi/100 ft)	Velocity (ft/sec)	Loss of Head (ft/100 ft)	Pressure Drop (psi/100 ft)	Velocity (ft/sec)	Loss of Head (ft/100 ft)	Pressure Drop (psi/100 ft)	Velocity (ft/sec)	Loss of Head (ft/100 ft)	Pressure Drop (psi/100 ft)
1,100	1.287	0.028	0.012	1.316	0.029	0.013	1.352	0.031	0.013	1.411	0.034	0.015	1.462	0.037	0.016	1.518	0.041	0.018	1.641	0.050	0.022
1,400	1.638	0.043	0.019	1.674	0.045	0.020	1.721	0.048	0.021	1.796	0.054	0.023	1.861	0.059	0.025	1.932	0.064	0.028	2.089	0.078	0.034
1,700	1.990	0.062	0.027	2.033	0.063	0.028	2.090	0.069	0.030	2.181	0.077	0.033	2.260	0.084	0.036	2.346	0.092	0.040	2.536	0.111	0.048
2,000	2.341	0.083	0.036	2.392	0.088	0.038	2.459	0.094	0.041	2.565	0.104	0.045	2.658	0.113	0.049	2.760	0.124	0.054	2.984	0.150	0.065
2,300	2.692	0.108	0.047	2.751	0.114	0.049	2.828	0.121	0.053	2.950	0.135	0.058	3.057	0.147	0.064	3.174	0.161	0.070	3.431	0.195	0.084
2,600	3.043	0.135	0.059	3.110	0.143	0.062	3.197	0.152	0.066	3.335	0.169	0.073	3.456	0.184	0.080	3.588	0.202	0.087	3.879	0.244	0.106
2,900	3.394	0.166	0.072	3.468	0.174	0.076	3.565	0.187	0.081	3.720	0.207	0.090	3.855	0.226	0.098	4.002	0.247	0.107	4.327	0.299	0.130
3,200	3.745	0.199	0.086	3.827	0.209	0.091	3.934	0.224	0.097	4.104	0.248	0.107	4.253	0.271	0.117	4.417	0.297	0.128	4.774	0.359	0.156
3,500	4.096	0.235	0.102	4.186	0.247	0.107	4.303	0.264	0.114	4.489	0.293	0.127	4.652	0.319	0.183	4.831	0.350	0.152	5.222	0.424	0.184
4,000	4.681	0.300	0.130	4.784	0.316	0.137	4.918	0.338	0.146	5.131	0.375	0.162	5.317	0.409	0.177	5.521	0.448	0.194	5.968	0.542	0.235
4,500	5.266	0.374	0.162	5.382	0.394	0.170	5.533	0.421	0.182	5.772	0.467	0.202	5.981	0.509	0.220	6.211	0.558	0.241	6.714	0.675	0.292
5,000	5.851	0.454	0.197	5.980	0.478	0.207	6.147	0.512	0.221	6.413	0.567	0.246	6.646	0.618	0.268	6.901	0.678	0.293	7.460	0.820	0.355
5,500	6.437	0.542	0.235	6.578	0.571	0.247	6.762	0.610	0.264	7.055	0.677	0.293	7.310	0.738	0.319	7.591	0.809	0.350	8.206	0.978	0.424
6,000	7.022	0.637	0.276	7.176	0.671	0.290	7.377	0.717	0.310	7.696	0.795	0.344	7.975	0.867	0.375	8.281	0.950	0.411	8.952	1.149	0.498
6,500	7.607	0.738	0.320	7.774	0.778	0.337	7.992	0.832	0.360	8.337	0.922	0.399	8.640	1.005	0.435	8.971	1.102	0.477	9.698	1.333	0.578
7,000	8.192	0.847	0.367	8.372	0.892	0.386	8.606	0.954	0.413	8.979	1.058	0.458	9.304	1.153	0.499	9.661	1.264	0.547	10.444	1.529	0.663
7,500	8.777	0.962	0.417	8.970	1.014	0.439	9.221	1.084	0.469	9.620	1.202	0.520	9.969	1.310	0.567	10.351	1.436	0.622	11.190	1.738	0.753
8,000	9.362	1.085	0.470	9.568	1.143	0.495	9.836	1.222	0.529	10.261	1.354	0.586	10.633	1.477	0.639	11.041	1.619	0.701	11.936	1.958	0.849

(continued on next page)

138 PVC PIPE—DESIGN AND INSTALLATION

Table B-2 Flow friction loss, AWWA C905 pipe (continued)

Flow (gpm)	20 in. CIOD (AWWA C905) DR 51 Actual OD 21.60 in. Pressure Rating 80 psi			20 in. CIOD (AWWA C905) DR 41 Actual OD 21.60 in. Pressure Rating 100 psi			20 in. CIOD (AWWA C905) DR 32.5 Actual OD 21.60 in. Pressure Rating 125 psi			20 in. CIOD (AWWA C905) DR 25 Actual OD 21.60 in. Pressure Rating 165 psi			20 in. CIOD (AWWA C905) DR 21 Actual OD 21.60 in. Pressure Rating 200 psi			20 in. CIOD (AWWA C905) DR 18 Actual OD 21.60 in. Pressure Rating 235 psi		
	Velocity (ft/sec)	Loss of Head (ft/100 ft)	Pressure Drop (psi/100 ft)	Velocity (ft/sec)	Loss of Head (ft/100 ft)	Pressure Drop (psi/100 ft)	Velocity (ft/sec)	Loss of Head (ft/100 ft)	Pressure Drop (psi/100 ft)	Velocity (ft/sec)	Loss of Head (ft/100 ft)	Pressure Drop (psi/100 ft)	Velocity (ft/sec)	Loss of Head (ft/100 ft)	Pressure Drop (psi/100 ft)	Velocity (ft/sec)	Loss of Head (ft/100 ft)	Pressure Drop (psi/100 ft)
1,400	1.336	0.026	0.011	1.365	0.028	0.012	1.403	0.029	0.013	1.464	0.033	0.014	1.517	0.036	0.015	1.575	0.039	0.017
1,900	1.813	0.046	0.020	1.852	0.048	0.021	1.904	0.052	0.022	1.986	0.057	0.025	2.058	0.063	0.027	2.137	0.069	0.030
2,400	2.290	0.071	0.031	2.339	0.075	0.032	2.405	0.080	0.035	2.509	0.089	0.038	2.600	0.097	0.042	2.700	0.106	0.046
2,900	2.767	0.101	0.044	2.827	0.106	0.046	2.906	0.113	0.049	3.032	0.126	0.054	3.142	0.137	0.059	3.262	0.150	0.065
3,400	3.244	0.135	0.059	3.314	0.142	0.062	3.407	0.152	0.066	3.554	0.169	0.073	3.683	0.184	0.080	3.825	0.202	0.087
3,900	3.721	0.174	0.076	3.801	0.184	0.079	3.908	0.196	0.085	4.077	0.218	0.094	4.225	0.237	0.103	4.387	0.260	0.113
4,400	4.198	0.218	0.095	4.289	0.230	0.099	4.409	0.246	0.106	4.600	0.272	0.118	4.766	0.297	0.128	4.950	0.325	0.141
4,900	4.675	0.266	0.115	4.776	0.280	0.121	4.910	0.300	0.130	5.122	0.332	0.144	5.308	0.362	0.157	5.512	0.397	0.172
5,400	5.152	0.319	0.138	5.263	0.335	0.145	5.411	0.359	0.155	5.645	0.398	0.172	5.850	0.434	0.188	6.075	0.475	0.206
6,000	5.724	0.387	0.168	5.848	0.408	0.176	6.013	0.436	0.189	6.272	0.483	0.209	6.500	0.527	0.228	6.655	0.578	0.250
7,000	6.678	0.515	0.223	6.823	0.542	0.235	7.015	0.580	0.251	7.318	0.643	0.278	7.583	0.701	0.304	7.875	0.769	0.333
8,000	7.632	0.660	0.286	7.798	0.694	0.301	8.017	0.743	0.322	8.363	0.823	0.356	8.666	0.898	0.389	9.000	0.984	0.426
9,000	8.586	0.821	0.356	8.772	0.864	0.374	9.019	0.924	0.400	9.408	1.024	0.443	9.749	1.117	0.483	10.124	1.224	0.530
10,000	9.540	0.997	0.432	9.747	1.050	0.454	10.021	1.123	0.486	10.454	1.245	0.539	10.833	1.357	0.588	11.249	1.488	0.644
11,000	10.494	1.190	0.516	10.722	1.252	0.542	11.023	1.340	0.580	11.499	1.485	0.643	11.916	1.619	0.701	12.374	1.775	0.768
12,000	11.448	1.398	0.606	11.696	1.472	0.637	12.025	1.574	0.681	12.544	1.745	0.755	12.999	1.903	0.824	13.499	2.086	0.903
13,000	12.402	1.621	0.703	12.671	1.707	0.739	13.028	1.826	0.790	13.590	2.023	0.867	14.083	2.207	0.955	14.624	2.419	1.047
14,000	13.356	1.860	0.806	13.646	1.958	0.847	14.030	2.094	0.907	14.635	2.321	1.005	15.166	2.531	1.096	15.749	2.775	1.201

(continued on next page)

APPENDIX B   139

Table B-2 Flow friction loss, AWWA C905 pipe (continued)

Flow (gpm)	24 in. CIOD (AWWA C905) DR 51 Actual OD 25.80 in. Pressure Rating 80 psi			24 in. CIOD (AWWA C905) DR 41 Actual OD 25.80 in. Pressure Rating 100 psi			24 in. CIOD (AWWA C905) DR 32.5 Actual OD 25.80 in. Pressure Rating 125 psi			24 in. CIOD (AWWA C905) DR 25 Actual OD 25.80 in. Pressure Rating 165 psi			24 in. CIOD (AWWA C905) DR 21 Actual OD 25.80 in. Pressure Rating 200 psi			24 in. CIOD (AWWA C905) DR 18 Actual OD 25.80 in. Pressure Rating 235 psi		
	Velocity (ft/sec)	Loss of Head (ft/100 ft)	Pressure Drop (psi/100 ft)	Velocity (ft/sec)	Loss of Head (ft/100 ft)	Pressure Drop (psi/100 ft)	Velocity (ft/sec)	Loss of Head (ft/100 ft)	Pressure Drop (psi/100 ft)	Velocity (ft/sec)	Loss of Head (ft/100 ft)	Pressure Drop (psi/100 ft)	Velocity (ft/sec)	Loss of Head (ft/100 ft)	Pressure Drop (psi/100 ft)	Velocity (ft/sec)	Loss of Head (ft/100 ft)	Pressure Drop (psi/100 ft)
2,000	1.337	0.021	0.009	1.366	0.022	0.010	1.405	0.024	0.010	1.465	0.027	0.012	1.519	0.029	0.013	1.577	0.032	0.014
2,500	1.672	0.032	0.014	1.708	0.034	0.015	1.756	0.036	0.016	1.832	0.040	0.017	1.898	0.044	0.019	1.971	0.048	0.021
3,000	2.006	0.045	0.020	2.049	0.048	0.021	2.107	0.051	0.022	2.198	0.056	0.024	2.278	0.061	0.027	2.365	0.067	0.029
3,500	2.340	0.060	0.026	2.391	0.063	0.027	2.458	0.068	0.029	2.564	0.075	0.032	2.657	0.082	0.035	2.760	0.090	0.039
4,000	2.675	0.077	0.033	2.732	0.081	0.035	2.809	0.087	0.038	2.931	0.096	0.042	3.037	0.105	0.045	3.154	0.115	0.050
4,500	3.009	0.096	0.042	3.074	0.101	0.044	3.161	0.108	0.047	3.297	0.119	0.052	3.417	0.130	0.056	3.548	0.143	0.062
5,000	3.343	0.116	0.050	3.416	0.122	0.053	3.512	0.131	0.057	3.664	0.145	0.063	3.796	0.158	0.069	3.942	0.174	0.075
5,500	3.678	0.139	0.060	3.757	0.146	0.063	3.863	0.156	0.068	4.030	0.173	0.075	4.176	0.189	0.082	4.336	0.207	0.090
6,000	4.012	0.163	0.071	4.099	0.172	0.074	4.214	0.184	0.080	4.396	0.204	0.088	4.556	0.222	0.096	4.731	0.243	0.105
6,500	4.346	0.189	0.082	4.440	0.199	0.086	4.565	0.213	0.092	4.763	0.236	0.102	4.935	0.257	0.111	5.125	0.282	0.122
7,500	5.015	0.247	0.107	5.123	0.260	0.112	5.268	0.278	0.120	5.495	0.308	0.133	5.695	0.336	0.145	5.913	0.368	0.159
8,500	5.684	0.311	0.135	5.806	0.327	0.142	5.970	0.350	0.152	6.228	0.388	0.168	6.454	0.423	0.183	6.702	0.464	0.201
9,500	6.352	0.382	0.166	6.490	0.402	0.174	6.672	0.430	0.186	6.961	0.477	0.206	7.213	0.520	0.225	7.491	0.570	0.247
10,500	7.021	0.460	0.199	7.173	0.484	0.209	7.375	0.518	0.224	7.693	0.574	0.248	7.972	0.626	0.271	8.279	0.686	0.297
11,500	7.689	0.544	0.236	7.856	0.573	0.248	8.077	0.613	0.265	8.426	0.679	0.294	8.732	0.741	0.321	9.067	0.812	0.351
12,500	8.358	0.635	0.275	8.539	0.668	0.289	8.779	0.715	0.310	9.159	0.793	0.343	9.491	0.864	0.374	9.856	0.947	0.410
13,500	9.027	0.732	0.317	9.222	0.771	0.334	9.482	0.825	0.357	9.891	0.914	0.396	10.251	0.997	0.432	10.644	1.093	0.473
14,500	9.695	0.836	0.362	9.905	0.880	0.381	10.184	0.941	0.407	10.624	1.043	0.452	11.010	1.138	0.493	11.432	1.247	0.540

(continued on next page)

140 PVC PIPE—DESIGN AND INSTALLATION

Table B-2 Flow friction loss, AWWA C905 pipe (continued)

Flow (gpm)	30 in. CIOD (AWWA C905) DR 51 Actual OD 32.00 in. Pressure Rating 80 psi			30 in. CIOD (AWWA C905) DR 41 Actual OD 32.00 in. Pressure Rating 100 psi			30 in. CIOD (AWWA C905) DR 32.5 Actual OD 32.00 in. Pressure Rating 125 psi			30 in. CIOD (AWWA C905) DR 25 Actual OD 32.00 in. Pressure Rating 165 psi			30 in. CIOD (AWWA C905) DR 21 Actual OD 32.00 in. Pressure Rating 200 psi		
	Velocity (ft/sec)	Loss of Head (ft/100 ft)	Pressure Drop (psi/100 ft)	Velocity (ft/sec)	Loss of Head (ft/100 ft)	Pressure Drop (psi/100 ft)	Velocity (ft/sec)	Loss of Head (ft/100 ft)	Pressure Drop (psi/100 ft)	Velocity (ft/sec)	Loss of Head (ft/100 ft)	Pressure Drop (psi/100 ft)	Velocity (ft/sec)	Loss of Head (ft/100 ft)	Pressure Drop (psi/100 ft)
3,000	1.304	0.016	0.007	1.332	0.017	0.007	1.370	0.018	0.008	1.429	0.020	0.009	1.481	0.022	0.009
3,500	1.521	0.021	0.009	1.554	0.022	0.010	1.598	0.024	0.010	1.667	0.026	0.011	1.727	0.029	0.012
4,000	1.738	0.027	0.012	1.776	0.028	0.012	1.826	0.030	0.013	1.905	0.034	0.015	1.974	0.037	0.016
4,500	1.956	0.034	0.015	1.998	0.035	0.015	2.055	0.038	0.016	2.143	0.042	0.018	2.221	0.046	0.020
5,000	2.173	0.041	0.018	2.220	0.043	0.019	2.283	0.046	0.020	2.381	0.051	0.022	2.468	0.056	0.024
5,500	2.390	0.049	0.021	2.442	0.051	0.022	2.511	0.055	0.024	2.620	0.061	0.026	2.715	0.066	0.029
6,000	2.608	0.057	0.025	2.664	0.060	0.026	2.739	0.064	0.028	2.858	0.071	0.031	2.961	0.078	0.034
6,500	2.825	0.066	0.029	2.886	0.070	0.030	2.968	0.075	0.032	3.096	0.083	0.036	3.208	0.090	0.039
7,000	3.042	0.076	0.033	3.108	0.080	0.035	3.196	0.086	0.037	3.334	0.095	0.041	3.455	0.104	0.045
8,000	3.477	0.097	0.042	3.552	0.103	0.044	3.653	0.110	0.048	3.810	0.122	0.053	3.949	0.133	0.057
10,000	4.346	0.147	0.064	4.440	0.155	0.067	4.566	0.166	0.072	4.763	0.184	0.080	4.936	0.201	0.087
12,000	5.215	0.206	0.090	5.328	0.217	0.094	5.479	0.233	0.101	5.716	0.258	0.112	5.923	0.281	0.122
14,000	6.085	0.275	0.119	6.217	0.289	0.125	6.392	0.309	0.134	6.668	0.343	0.148	6.910	0.374	0.162
16,000	6.954	0.352	0.152	7.105	0.370	0.160	7.305	0.396	0.171	7.621	0.439	0.190	7.897	0.479	0.207
18,000	7.823	0.438	0.190	7.993	0.460	0.199	8.218	0.493	0.213	8.573	0.546	0.236	8.884	0.596	0.258
20,000	8.692	0.532	0.231	8.881	0.560	0.242	9.131	0.599	0.259	9.526	0.664	0.287	9.871	0.724	0.313
22,000	9.561	0.634	0.275	9.769	0.668	0.289	10.044	0.715	0.309	10.478	0.792	0.343	10.858	0.864	0.374
24,000	10.431	0.745	0.323	10.657	0.785	0.340	10.958	0.839	0.363	11.431	0.930	0.403	11.846	1.015	0.439

(continued on next page)

Table B-2 Flow friction loss, AWWA C905 pipe (continued)

Flow (gpm)	36 in. CIOD (AWWA C905) DR 51 Actual OD 38.30 in. Pressure Rating 80 psi			36 in. CIOD (AWWA C905) DR 41 Actual OD 38.30 in. Pressure Rating 100 psi			36 in. CIOD (AWWA C905) DR 32.5 Actual OD 38.30 in. Pressure Rating 125 psi			36 in. CIOD (AWWA C905) DR 25 Actual OD 38.30 in. Pressure Rating 165 psi			36 in. CIOD (AWWA C905) DR 21 Actual OD 38.30 in. Pressure Rating 200 psi		
	Velocity (ft/sec)	Loss of Head (ft/100 ft)	Pressure Drop (psi/100 ft)	Velocity (ft/sec)	Loss of Head (ft/100 ft)	Pressure Drop (psi/100 ft)	Velocity (ft/sec)	Loss of Head (ft/100 ft)	Pressure Drop (psi/100 ft)	Velocity (ft/sec)	Loss of Head (ft/100 ft)	Pressure Drop (psi/100 ft)	Velocity (ft/sec)	Loss of Head (ft/100 ft)	Pressure Drop (psi/100 ft)
5,000	1.517	0.017	0.007	1.550	0.018	0.008	1.593	0.019	0.008	1.662	0.021	0.009	1.723	0.023	0.010
5,500	1.669	0.020	0.009	1.705	0.021	0.009	1.753	0.023	0.010	1.829	0.025	0.011	1.895	0.028	0.012
6,000	1.820	0.024	0.010	1.860	0.025	0.011	1.912	0.027	0.012	1.995	0.030	0.013	2.067	0.032	0.014
6,500	1.972	0.028	0.012	2.015	0.029	0.013	2.071	0.031	0.013	2.161	0.035	0.015	2.240	0.038	0.016
7,000	2.124	0.032	0.014	2.170	0.033	0.014	2.231	0.036	0.015	2.327	0.040	0.017	2.412	0.043	0.019
8,000	2.427	0.041	0.018	2.480	0.043	0.019	2.549	0.046	0.020	2.660	0.051	0.022	2.756	0.055	0.024
11,000	3.337	0.073	0.032	3.410	0.077	0.033	3.505	0.083	0.036	3.657	0.092	0.040	3.791	0.100	0.043
14,000	4.248	0.115	0.050	4.340	0.121	0.052	4.462	0.129	0.056	4.655	0.143	0.062	4.824	0.156	0.068
17,000	5.158	0.164	0.071	5.270	0.173	0.075	5.418	0.185	0.080	5.652	0.205	0.089	5.857	0.223	0.097
20,000	6.068	0.222	0.096	6.200	0.234	0.101	6.374	0.250	0.108	6.650	0.277	0.120	6.891	0.302	0.131
23,000	6.978	0.287	0.125	7.130	0.302	0.131	7.330	0.324	0.140	7.647	0.359	0.155	7.924	0.391	0.169
26,000	7.889	0.361	0.156	8.060	0.380	0.164	8.286	0.406	0.176	8.645	0.450	0.195	8.958	0.491	0.212
29,000	8.799	0.441	0.191	8.990	0.465	0.201	9.242	0.497	0.215	9.642	0.551	0.239	9.992	0.601	0.260
32,000	9.709	0.530	0.230	9.920	0.558	0.241	10.198	0.596	0.258	10.639	0.661	0.286	11.025	0.721	0.312
35,000	10.619	0.625	0.271	10.850	0.658	0.285	11.154	0.704	0.305	11.637	0.781	0.338	12.059	0.851	0.368
38,000	11.529	0.728	0.316	11.780	0.767	0.332	12.110	0.820	0.355	12.634	0.909	0.393	13.093	0.991	0.429
41,000	12.440	0.838	0.363	12.710	0.882	0.382	13.066	0.944	0.409	13.632	1.046	0.453	14.126	1.141	0.494
44,000	13.350	0.955	0.414	13.640	1.006	0.435	14.022	1.076	0.466	14.629	1.192	0.516	15.160	1.300	0.563

(continued on next page)

Table B-2 Flow friction loss, AWWA C905 pipe (continued)

Flow (gpm)	42 in. CIOD (AWWA C905) DR 51 Actual OD 44.50 in. Pressure Rating 80 psi			42 in. CIOD (AWWA C905) DR 41 Actual OD 44.50 in. Pressure Rating 100 psi			42 in. CIOD (AWWA C905) DR 32.5 Actual OD 44.50 in. Pressure Rating 125 psi			42 in. CIOD (AWWA C905) DR 25 Actual OD 44.50 in. Pressure Rating 165 psi		
	Velocity (ft/sec)	Loss of Head (ft/100 ft)	Pressure Drop (psi/100 ft)	Velocity (ft/sec)	Loss of Head (ft/100 ft)	Pressure Drop (psi/100 ft)	Velocity (ft/sec)	Loss of Head (ft/100 ft)	Pressure Drop (psi/100 ft)	Velocity (ft/sec)	Loss of Head (ft/100 ft)	Pressure Drop (psi/100 ft)
7,000	1.573	0.015	0.007	1.607	0.016	0.007	1.653	0.017	0.007	1.724	0.019	0.008
8,000	1.798	0.020	0.008	1.837	0.021	0.009	1.889	0.022	0.010	1.970	0.024	0.011
9,000	2.023	0.024	0.011	2.067	0.026	0.011	2.125	0.027	0.012	2.217	0.030	0.013
12,000	2.697	0.041	0.018	2.755	0.044	0.019	2.833	0.047	0.020	2.955	0.052	0.022
15,000	3.371	0.063	0.027	3.444	0.066	0.029	3.541	0.071	0.031	3.694	0.078	0.034
18,000	4.045	0.088	0.038	4.133	0.093	0.040	4.249	0.099	0.043	4.433	0.110	0.048
21,000	4.720	0.117	0.051	4.822	0.123	0.053	4.958	0.132	0.057	5.172	0.146	0.063
24,000	5.394	0.150	0.065	5.511	0.158	0.068	5.666	0.169	0.073	5.911	0.187	0.081
27,000	6.068	0.186	0.081	6.200	0.196	0.085	6.374	0.210	0.091	6.650	0.233	0.101
30,000	6.742	0.226	0.098	6.889	0.239	0.103	7.082	0.255	0.111	7.389	0.283	0.123
33,000	7.416	0.270	0.117	7.578	0.285	0.123	7.790	0.305	0.132	8.128	0.338	0.146
36,000	8.091	0.317	0.137	8.266	0.335	0.145	8.499	0.358	0.155	8.866	0.397	0.172
39,000	8.765	0.368	0.159	8.955	0.388	0.168	9.207	0.415	0.180	9.605	0.460	0.199
42,000	9.439	0.422	0.183	9.644	0.445	0.193	9.915	0.476	0.206	10.344	0.528	0.229
45,000	10.113	0.480	0.208	10.333	0.506	0.219	10.623	0.541	0.234	11.083	0.600	0.260
48,000	10.787	0.541	0.234	11.022	0.570	0.247	11.331	0.610	0.264	11.822	0.676	0.293
51,000	11.462	0.605	0.262	11.711	0.638	0.276	12.040	0.682	0.295	12.561	0.756	0.327

(continued on next page)

Table B-2 Flow friction loss, AWWA C905 pipe (continued)

Flow (gpm)	48 in. CIOD (AWWA C905) DR 51 Actual OD 50.80 in. Pressure Rating 80 psi			48 in. CIOD (AWWA C905) DR 41 Actual OD 50.80 in. Pressure Rating 100 psi			48 in. CIOD (AWWA C905) DR 32.5 Actual OD 50.80 in. Pressure Rating 125 psi			48 in. CIOD (AWWA C905) DR 25 Actual OD 50.80 in. Pressure Rating 165 psi		
	Velocity (ft/sec)	Loss of Head (ft/100 ft)	Pressure Drop (psi/100 ft)	Velocity (ft/sec)	Loss of Head (ft/100 ft)	Pressure Drop (psi/100 ft)	Velocity (ft/sec)	Loss of Head (ft/100 ft)	Pressure Drop (psi/100 ft)	Velocity (ft/sec)	Loss of Head (ft/100 ft)	Pressure Drop (psi/100 ft)
8,000	1.380	0.010	0.004	1.410	0.011	0.005	1.449	0.012	0.005	1.512	0.013	0.006
10,000	1.725	0.016	0.007	1.762	0.016	0.007	1.812	0.018	0.008	1.890	0.019	0.008
15,000	2.587	0.033	0.014	2.643	0.035	0.015	2.717	0.037	0.016	2.835	0.041	0.018
20,000	3.449	0.056	0.024	3.524	0.059	0.026	3.623	0.063	0.027	3.780	0.070	0.030
25,000	4.311	0.085	0.037	4.405	0.089	0.039	4.529	0.096	0.041	4.725	0.106	0.046
30,000	5.174	0.119	0.052	5.286	0.125	0.054	5.435	0.134	0.058	5.670	0.149	0.064
35,000	6.036	0.158	0.069	6.167	0.167	0.072	6.341	0.178	0.077	6.615	0.198	0.086
40,000	6.898	0.203	0.088	7.048	0.213	0.092	7.246	0.228	0.099	7.560	0.253	0.110
45,000	7.761	0.252	0.109	7.929	0.265	0.115	8.152	0.284	0.123	8.504	0.315	0.136
50,000	8.623	0.306	0.133	8.810	0.323	0.140	9.058	0.345	0.150	9.449	0.383	0.166
55,000	9.485	0.365	0.158	9.691	0.385	0.167	9.964	0.412	0.178	10.394	0.457	0.198
60,000	10.348	0.429	0.186	10.572	0.452	0.196	10.870	0.484	0.210	11.339	0.536	0.232
65,000	11.210	0.498	0.216	11.453	0.525	0.227	11.775	0.561	0.243	12.284	0.622	0.269
70,000	12.072	0.571	0.247	12.334	0.602	0.261	12.681	0.644	0.279	13.229	0.714	0.309

144 PVC PIPE—DESIGN AND INSTALLATION

Table B-3 Flow friction loss, ASTM D2241/AWWA C905 pipe

Flow (gpm)	14 in. I.P.S. OD (ASTM D2241/AWWA C905) DR 41 Actual OD 14.00 in. Pressure Rating 100 psi			14 in. I.P.S. OD (ASTM D2241/AWWA C905) DR 32.5 Actual OD 14.00 in. Pressure Rating 125 psi			14 in. I.P.S. OD (ASTM D2241/AWWA C905) DR 26 Actual OD 14.00 in. Pressure Rating 160 psi			14 in. I.P.S. OD (ASTM D2241/AWWA C905) DR 21 Actual OD 14.00 in. Pressure Rating 200 psi		
	Velocity (ft/sec)	Loss of Head (ft/100 ft)	Pressure Drop (psi/100 ft)	Velocity (ft/sec)	Loss of Head (ft/100 ft)	Pressure Drop (psi/100 ft)	Velocity (ft/sec)	Loss of Head (ft/100 ft)	Pressure Drop (psi/100 ft)	Velocity (ft/sec)	Loss of Head (ft/100 ft)	Pressure Drop (psi/100 ft)
700	1.624	0.063	0.027	1.669	0.067	0.029	1.729	0.073	0.032	1.805	0.081	0.035
800	1.856	0.080	0.035	1.908	0.086	0.037	1.976	0.094	0.041	2.062	0.104	0.045
900	2.088	0.100	0.043	2.146	0.107	0.046	2.223	0.117	0.050	2.320	0.129	0.056
1,000	2.320	0.122	0.053	2.385	0.130	0.056	2.470	0.142	0.061	2.578	0.157	0.068
1,100	2.552	0.145	0.063	2.623	0.155	0.067	2.717	0.169	0.073	2.836	0.188	0.081
1,200	2.784	0.171	0.074	2.862	0.182	0.079	2.964	0.199	0.086	3.094	0.220	0.095
1,300	3.016	0.198	0.086	3.100	0.212	0.092	3.211	0.230	0.100	3.351	0.256	0.111
1,500	3.480	0.258	0.112	3.577	0.276	0.119	3.705	0.300	0.130	3.867	0.333	0.144
1,700	3.944	0.325	0.141	4.054	0.348	0.151	4.199	0.379	0.164	4.383	0.420	0.182
1,900	4.408	0.399	0.173	4.531	0.427	0.185	4.693	0.465	0.201	4.898	0.516	0.224
2,100	4.872	0.481	0.208	5.008	0.514	0.223	5.187	0.560	0.242	5.414	0.622	0.269
2,300	5.336	0.569	0.246	5.485	0.609	0.263	5.681	0.663	0.287	5.929	0.736	0.318
2,500	5.800	0.664	0.287	5.962	0.710	0.307	6.175	0.774	0.335	6.445	0.858	0.372
3,000	6.959	0.931	0.403	7.154	0.996	0.431	7.410	1.084	0.469	7.734	1.203	0.521
3,500	8.119	1.238	0.536	8.347	1.324	0.573	8.645	1.443	0.625	9.023	1.601	0.693
4,000	9.279	1.586	0.686	9.539	1.696	0.734	9.880	1.847	0.800	10.312	2.050	0.887
4,500	10.439	1.972	0.854	10.731	2.109	0.913	11.115	2.298	0.995	11.601	2.550	1.104
5,000	11.599	2.397	1.038	11.924	2.564	1.110	12.350	2.793	1.209	12.890	3.099	1.342

(continued on next page)

Table B-3   Flow friction loss, ASTM D2241/AWWA C905 pipe (continued)

Flow (gpm)	16 in. I.P.S. OD (ASTM D2241/AWWA C905) DR 41 Actual OD 16.00 in. Pressure Rating 100 psi			16 in. I.P.S. OD (ASTM D2241/AWWA C905) DR 32.5 Actual OD 16.00 in. Pressure Rating 125 psi			16 in. I.P.S. OD (ASTM D2241/AWWA C905) DR 26 Actual OD 16.00 in. Pressure Rating 160 psi			16 in. I.P.S. OD (ASTM D2241/AWWA C905) DR 21 Actual OD 16.00 in. Pressure Rating 200 psi		
	Velocity (ft/sec)	Loss of Head (ft/100 ft)	Pressure Drop (psi/100 ft)	Velocity (ft/sec)	Loss of Head (ft/100 ft)	Pressure Drop (psi/100 ft)	Velocity (ft/sec)	Loss of Head (ft/100 ft)	Pressure Drop (psi/100 ft)	Velocity (ft/sec)	Loss of Head (ft/100 ft)	Pressure Drop (psi/100 ft)
900	1.599	0.052	0.023	1.643	0.056	0.024	1.702	0.061	0.006	1.777	0.068	0.029
1,000	1.776	0.064	0.028	1.826	0.068	0.029	1.891	0.074	0.032	1.974	0.082	0.036
1,100	1.954	0.076	0.033	2.009	0.081	0.035	2.081	0.088	0.038	2.172	0.098	0.042
1,200	2.132	0.089	0.039	2.191	0.095	0.041	2.270	0.104	0.045	2.369	0.115	0.050
1,300	2.309	0.103	0.045	2.374	0.110	0.048	2.459	0.120	0.052	2.566	0.134	0.058
1,400	2.487	0.119	0.051	2.556	0.127	0.055	2.648	0.138	0.060	2.764	0.153	0.066
1,500	2.664	0.135	0.058	2.739	0.144	0.062	2.837	0.157	0.068	2.961	0.174	0.075
1,700	3.020	0.170	0.074	3.104	0.182	0.079	3.215	0.198	0.086	3.356	0.220	0.095
1,900	3.375	0.209	0.090	3.469	0.223	0.097	3.594	0.243	0.105	3.751	0.270	0.117
2,100	3.730	0.251	0.109	3.835	0.269	0.116	3.972	0.293	0.127	4.146	0.325	0.141
2,300	4.085	0.297	0.129	4.200	0.318	0.138	4.350	0.346	0.150	4.540	0.384	0.166
2,500	4.441	0.347	0.150	4.565	0.371	0.161	4.729	0.404	0.175	4.935	0.448	0.194
3,000	5.329	0.486	0.210	5.478	0.520	0.225	5.674	0.566	0.245	5.922	0.629	0.272
3,500	6.217	0.647	0.280	6.391	0.692	0.299	6.620	0.754	0.326	6.909	0.836	0.362
4,000	7.105	0.828	0.359	7.304	0.886	0.383	7.566	0.965	0.418	7.896	1.071	0.464
4,500	7.993	1.030	0.446	8.217	1.102	0.477	8.512	1.200	0.520	8.884	1.332	0.577
5,000	8.881	1.252	0.542	9.130	1.339	0.580	9.457	1.459	0.632	9.871	1.619	0.701
5,500	9.769	1.494	0.647	10.043	1.598	0.692	10.403	1.741	0.754	10.858	1.932	0.836

(continued on next page)

146 PVC PIPE DESIGN AND INSTALLATION

Table B-3   Flow friction loss, ASTM D2241/AWWA C905 pipe (continued)

Flow (gpm)	18 in. I.P.S. OD (ASTM D2241/AWWA C905) DR 41 Actual OD 18.00 in. Pressure Rating 100 psi			18 in. I.P.S. OD (ASTM D2241/AWWA C905) DR 32.5 Actual OD 18.00 in. Pressure Rating 125 psi			18 in. I.P.S. OD (ASTM D2241/AWWA C905) DR 26 Actual OD 18.00 in. Pressure Rating 160 psi			18 in. I.P.S. OD (ASTM D2241/AWWA C905) DR 21 Actual OD 18.00 in. Pressure Rating 200 psi		
	Velocity (ft/sec)	Loss of Head (ft/100 ft)	Pressure Drop (psi/100 ft)	Velocity (ft/sec)	Loss of Head (ft/100 ft)	Pressure Drop (psi/100 ft)	Velocity (ft/sec)	Loss of Head (ft/100 ft)	Pressure Drop (psi/100 ft)	Velocity (ft/sec)	Loss of Head (ft/100 ft)	Pressure Drop (psi/100 ft)
900	1.263	0.029	0.013	1.299	0.032	0.014	1.345	0.034	0.015	1.404	0.038	0.016
1,200	1.684	0.050	0.022	1.731	0.054	0.023	1.793	0.059	0.025	1.872	0.065	0.028
1,500	2.105	0.076	0.033	2.164	0.081	0.035	2.242	0.088	0.038	2.340	0.098	0.042
1,800	2.526	0.106	0.046	2.597	0.114	0.049	2.690	0.124	0.054	2.808	0.138	0.060
2,100	2.947	0.142	0.061	3.030	0.151	0.066	3.138	0.165	0.071	3.276	0.183	0.079
2,400	3.368	0.181	0.078	3.163	0.194	0.084	3.587	0.211	0.091	3.744	0.234	0.101
2,700	3.789	0.225	0.098	3.896	0.241	0.104	4.035	0.263	0.114	4.212	0.292	0.126
3,000	4.210	0.274	0.119	4.329	0.293	0.127	4.483	0.319	0.138	4.679	0.354	0.153
3,300	4.631	0.327	0.142	4.762	0.350	0.151	4.932	0.381	0.165	5.147	0.423	0.183
3,500	4.912	0.365	0.158	5.050	0.390	0.169	5.231	0.425	0.184	5.459	0.472	0.204
4,000	5.614	0.467	0.202	5.772	0.500	0.216	5.978	0.544	0.236	6.239	0.604	0.261
4,500	6.315	0.581	0.251	6.493	0.621	0.269	6.725	0.677	0.293	7.019	0.751	0.325
5,000	7.017	0.706	0.306	7.214	0.755	0.327	7.472	0.823	0.356	7.799	0.913	0.395
5,500	7.719	0.842	0.365	7.936	0.901	0.390	8.220	0.981	0.425	8.579	1.089	0.471
6,000	8.420	0.989	0.428	8.657	1.058	0.458	8.967	1.153	0.499	9.359	1.279	0.554
6,500	9.122	1.148	0.497	9.379	1.228	0.531	9.714	1.337	0.579	10.139	1.484	0.642
7,000	9.824	1.316	0.570	10.100	1.408	0.610	10.461	1.534	0.664	10.919	1.702	0.737
7,500	10.526	1.496	0.648	10.822	1.600	0.693	11.209	1.743	0.755	11.699	1.934	0.837

(continued on next page)

Table B-3 Flow friction loss, ASTM D2241/AWWA C905 pipe (continued)

Flow (gpm)	20 in. I.P.S. OD (ASTM D2241/AWWA C905) DR 41 Actual OD 20.00 in. Pressure Rating 100 psi			20 in. I.P.S. OD (ASTM D2241/AWWA C905) DR 32.5 Actual OD 20.00 in. Pressure Rating 125 psi			20 in. I.P.S. OD (ASTM D2241/AWWA C905) DR 26 Actual OD 20.00 in. Pressure Rating 160 psi			20 in. I.P.S. OD (ASTM D2241/AWWA C905) DR 21 Actual OD 20.00 in. Pressure Rating 200 psi		
	Velocity (ft/sec)	Loss of Head (ft/100 ft)	Pressure Drop (psi/100 ft)	Velocity (ft/sec)	Loss of Head (ft/100 ft)	Pressure Drop (psi/100 ft)	Velocity (ft/sec)	Loss of Head (ft/100 ft)	Pressure Drop (psi/100 ft)	Velocity (ft/sec)	Loss of Head (ft/100 ft)	Pressure Drop (psi/100 ft)
1,100	1.251	0.026	0.011	1.286	0.027	0.012	1.332	0.030	0.013	1.390	0.033	0.014
1,400	1.592	0.040	0.017	1.636	0.043	0.019	1.695	0.047	0.020	1.769	0.052	0.022
1,700	1.933	0.057	0.025	1.987	0.061	0.027	2.058	0.067	0.029	2.148	0.074	0.032
2,000	2.274	0.077	0.034	2.337	0.083	0.036	2.421	0.090	0.039	2.527	0.100	0.043
2,300	2.615	0.100	0.043	2.688	0.107	0.046	2.784	0.117	0.051	2.906	0.130	0.056
2,600	2.956	0.126	0.055	3.039	0.135	0.058	3.147	0.147	0.064	3.285	0.163	0.070
2,900	3.297	0.154	0.067	3.389	0.165	0.071	3.511	0.180	0.078	3.664	0.199	0.086
3,200	3.638	0.185	0.080	3.740	0.198	0.086	3.874	0.216	0.093	4.043	0.239	0.104
3,500	3.979	0.218	0.095	4.090	0.234	0.101	4.237	0.254	0.110	4.422	0.282	0.122
4,000	4.547	0.280	0.121	4.675	0.299	0.130	4.842	0.326	0.141	5.053	0.362	0.157
4,500	5.116	0.348	0.151	5.259	0.372	0.161	5.447	0.405	0.175	5.685	0.450	0.195
5,000	5.684	0.423	0.183	5.843	0.452	0.196	6.053	0.493	0.213	6.317	0.547	0.237
5,500	6.253	0.504	0.218	6.428	0.540	0.234	6.658	0.588	0.254	6.948	0.652	0.282
6,000	6.821	0.593	0.257	7.012	0.634	0.274	7.263	0.691	0.299	7.580	0.766	0.332
6,500	7.389	0.687	0.298	7.596	0.735	0.318	7.868	0.801	0.347	8.212	0.889	0.385
7,000	7.958	0.789	0.341	8.181	0.843	0.365	8.474	0.919	0.398	8.843	1.019	0.441
7,500	8.526	0.896	0.388	8.765	0.958	0.415	9.079	1.044	0.452	9.475	1.158	0.501
8,000	9.095	1.010	0.437	9.349	1.080	0.468	9.684	1.176	0.509	10.107	1.305	0.565

(continued on next page)

Table B-3 Flow friction loss, ASTM D2241/AWWA C905 pipe (continued)

Flow (gpm)	24 in. I.P.S. OD (ASTM D2241/AWWA C905) DR 41 Actual OD 24.00 in. Pressure Rating 100 psi			24 in. I.P.S. OD (ASTM D2241/AWWA C905) DR 32.5 Actual OD 24.00 in. Pressure Rating 125 psi			24 in. I.P.S. OD (ASTM D2241/AWWA C905) DR 26 Actual OD 24.00 in. Pressure Rating 160 psi			24 in. I.P.S. OD (ASTM D2241/AWWA C905) DR 21 Actual OD 24.00 in. Pressure Rating 200 psi		
	Velocity (ft/sec)	Loss of Head (ft/100 ft)	Pressure Drop (psi/100 ft)	Velocity (ft/sec)	Loss of Head (ft/100 ft)	Pressure Drop (psi/100 ft)	Velocity (ft/sec)	Loss of Head (ft/100 ft)	Pressure Drop (psi/100 ft)	Velocity (ft/sec)	Loss of Head (ft/100 ft)	Pressure Drop (psi/100 ft)
1,600	1.263	0.021	0.009	1.298	0.023	0.010	1.345	0.025	0.011	1.404	0.027	0.012
1,900	1.500	0.029	0.013	1.542	0.031	0.013	1.597	0.034	0.015	1.667	0.038	0.016
2,200	1.737	0.038	0.016	1.785	0.041	0.018	1.850	0.044	0.019	1.930	0.049	0.021
2,500	1.974	0.048	0.021	2.029	0.052	0.022	2.102	0.056	0.024	2.194	0.062	0.027
2,800	2.210	0.059	0.026	2.272	0.064	0.028	2.354	0.069	0.030	2.457	0.077	0.033
3,100	2.447	0.072	0.031	2.516	0.077	0.033	2.606	0.084	0.036	2.720	0.093	0.040
3,400	2.684	0.085	0.037	2.759	0.091	0.039	2.858	0.099	0.043	2.983	0.110	0.048
3,700	2.921	0.100	0.043	3.003	0.107	0.046	3.111	0.116	0.050	3.246	0.129	0.056
4,000	3.158	0.115	0.050	3.246	0.123	0.053	3.363	0.134	0.058	3.510	0.149	0.064
4,500	3.552	0.143	0.062	3.652	0.153	0.066	3.783	0.167	0.072	3.948	0.185	0.080
5,000	3.947	0.174	0.075	4.058	0.186	0.081	4.204	0.203	0.088	4.387	0.225	0.097
5,500	4.342	0.208	0.090	4.463	0.222	0.096	4.624	0.242	0.105	4.826	0.269	0.116
6,000	4.736	0.244	0.106	4.869	0.261	0.113	5.044	0.284	0.123	5.265	0.316	0.137
6,500	5.131	0.283	0.123	5.275	0.303	0.131	5.465	0.330	0.143	5.703	0.366	0.158
7,000	5.526	0.325	0.141	5.681	0.347	0.150	5.885	0.378	0.164	6.142	0.420	0.182
7,500	5.921	0.369	0.160	6.086	0.395	0.171	6.305	0.430	0.186	6.581	0.477	0.207
8,000	6.315	0.416	0.180	6.492	0.445	0.193	6.726	0.485	0.210	7.019	0.538	0.233
8,500	6.710	0.465	0.201	6.898	0.498	0.215	7.146	0.542	0.235	7.458	0.602	0.260

(continued on next page)

Table B-3 Flow friction loss, ASTM D2241/AWWA C905 pipe (continued)

Flow (gpm)	30 in. I.P.S. OD (ASTM D2241/AWWA C905) DR 41 Actual OD 30.00 in. Pressure Rating 100 psi			30 in. I.P.S. OD (ASTM D2241/AWWA C905) DR 32.5 Actual OD 30.00 in. Pressure Rating 125 psi			30 in. I.P.S. OD (ASTM D2241/AWWA C905) DR 26 Actual OD 30.00 in. Pressure Rating 160 psi			30 in. I.P.S. OD (ASTM D2241/AWWA C905) DR 21 Actual OD 30.00 in. Pressure Rating 200 psi		
	Velocity (ft/sec)	Loss of Head (ft/100 ft)	Pressure Drop (psi/100 ft)	Velocity (ft/sec)	Loss of Head (ft/100 ft)	Pressure Drop (psi/100 ft)	Velocity (ft/sec)	Loss of Head (ft/100 ft)	Pressure Drop (psi/100 ft)	Velocity (ft/sec)	Loss of Head (ft/100 ft)	Pressure Drop (psi/100 ft)
2,600	1.314	0.018	0.008	1.351	0.019	0.008	1.399	0.020	0.009	1.460	0.023	0.010
2,900	1.465	0.021	0.009	1.506	0.023	0.010	1.560	0.025	0.011	1.628	0.028	0.012
3,200	1.617	0.026	0.011	1.662	0.028	0.012	1.722	0.030	0.013	1.797	0.033	0.014
3,500	1.768	0.030	0.013	1.818	0.032	0.014	1.833	0.035	0.015	1.965	0.039	0.017
3,800	1.920	0.035	0.015	1.974	0.038	0.016	2.045	0.041	0.018	2.134	0.046	0.020
4,100	2.072	0.041	0.018	2.130	0.044	0.019	2.206	0.047	0.021	2.302	0.053	0.023
4,400	2.223	0.046	0.020	2.286	0.050	0.021	2.367	0.054	0.023	2.471	0.060	0.026
4,700	2.375	0.052	0.023	2.441	0.056	0.024	2.529	0.061	0.026	2.639	0.068	0.029
5,000	2.526	0.059	0.025	2.597	0.063	0.027	2.690	0.069	0.030	2.807	0.076	0.033
6,000	3.032	0.082	0.036	3.117	0.088	0.038	3.228	0.096	0.042	3.369	0.107	0.046
8,000	4.042	0.140	0.061	4.156	0.150	0.065	4.304	0.164	0.071	4.492	0.182	0.079
10,000	5.053	0.212	0.092	5.194	0.227	0.098	5.381	0.247	0.107	5.615	0.274	0.119
12,000	6.063	0.298	0.129	6.233	0.318	0.138	6.457	0.347	0.150	6.738	0.385	0.166
14,000	7.074	0.396	0.171	7.272	0.423	0.183	7.533	0.461	0.200	7.861	0.512	0.222
16,000	8.084	0.507	0.219	8.311	0.542	0.235	8.609	0.591	0.256	8.984	0.655	0.284
18,000	9.095	0.630	0.273	9.350	0.674	0.292	9.685	0.735	0.318	10.107	0.815	0.353
20,000	10.105	0.766	0.332	10.389	0.820	0.355	10.761	0.893	0.387	11.230	0.991	0.429
22,000	11.116	0.914	0.396	11.428	0.978	0.423	11.837	1.065	0.461	12.353	1.182	0.512

(continued on next page)

150 PVC PIPE—DESIGN AND INSTALLATION

Table B-3 Flow friction loss, ASTM D2241/AWWA C905 pipe (continued)

Flow (gpm)	36 in. I.P.S. OD (ASTM D2241/AWWA C905) DR 41 Actual OD 36.00 in. Pressure Rating 100 psi			36 in. I.P.S. OD (ASTM D2241/AWWA C905) DR 32.5 Actual OD 36.00 in. Pressure Rating 125 psi			36 in. I.P.S. OD (ASTM D2241/AWWA C905) DR 26 Actual OD 36.00 in. Pressure Rating 160 psi			36 in. I.P.S. OD (ASTM D2241/AWWA C905) DR 21 Actual OD 36.00 in. Pressure Rating 200 psi		
	Velocity (ft/sec)	Loss of Head (ft/100 ft)	Pressure Drop (psi/100 ft)	Velocity (ft/sec)	Loss of Head (ft/100 ft)	Pressure Drop (psi/100 ft)	Velocity (ft/sec)	Loss of Head (ft/100 ft)	Pressure Drop (psi/100 ft)	Velocity (ft/sec)	Loss of Head (ft/100 ft)	Pressure Drop (psi/100 ft)
4,000	1.403	0.016	0.007	1.443	0.017	0.007	1.495	0.019	0.008	1.560	0.021	0.009
5,000	1.754	0.024	0.010	1.804	0.026	0.011	1.868	0.028	0.012	1.950	0.031	0.014
6,000	2.105	0.034	0.015	2.164	0.036	0.016	2.242	0.040	0.017	2.340	0.044	0.019
7,000	2.456	0.045	0.020	2.525	0.048	0.021	2.616	0.053	0.023	2.730	0.058	0.025
8,000	2.807	0.058	0.025	2.886	0.062	0.027	2.989	0.067	0.029	3.119	0.075	0.032
9,000	3.158	0.072	0.031	3.247	0.077	0.033	3.363	0.084	0.036	3.509	0.093	0.040
10,000	3.509	0.087	0.038	3.607	0.094	0.040	3.737	0.102	0.044	3.899	0.113	0.049
11,000	3.860	0.104	0.045	3.968	0.112	0.048	4.110	0.122	0.053	4.289	0.135	0.058
12,000	4.210	0.123	0.053	4.329	0.131	0.057	4.484	0.143	0.062	4.679	0.158	0.069
13,000	4.561	0.142	0.062	4.690	0.152	0.066	4.858	0.166	0.072	5.069	0.184	0.080
15,000	5.263	0.185	0.080	5.411	0.198	0.086	5.605	0.216	0.093	5.849	0.239	0.104
17,000	5.965	0.234	0.101	6.133	0.250	0.108	6.352	0.272	0.118	6.629	0.302	0.131
19,000	6.667	0.287	0.124	6.854	0.307	0.133	7.100	0.334	0.145	7.409	0.371	0.161
21,000	7.368	0.345	0.150	7.576	0.370	0.160	7.847	0.403	0.174	8.189	0.447	0.193
23,000	8.070	0.409	0.177	8.297	0.437	0.189	8.594	0.476	0.206	8.968	0.529	0.229
25,000	8.772	0.477	0.207	9.019	0.510	0.221	9.342	0.556	0.241	9.748	0.617	0.267
27,000	9.474	0.550	0.238	9.740	0.589	0.255	10.089	0.641	0.278	10.528	0.711	0.308
29,000	10.175	0.628	0.272	10.462	0.672	0.291	10.836	0.732	0.317	11.308	0.812	0.351

Table B-4 Flow friction loss, AWWA C909 PVCO pipe

Flow (gpm)	4 in. CIOD (AWWA C909) CL 100			4 in. CIOD (AWWA C909) CL 150			4 in. CIOD (AWWA C909) CL 200		
	Velocity (ft/sec)	Loss of Head (ft/100 ft)	Pressure Drop (psi/100 ft)	Velocity (ft/sec)	Loss of Head (ft/100 ft)	Pressure Drop (psi/100 ft)	Velocity (ft/sec)	Loss of Head (ft/100 ft)	Pressure Drop (psi/100 ft)
20	0.390	0.015	0.007	0.405	0.017	0.007	0.422	0.019	0.008
25	0.487	0.023	0.010	0.506	0.026	0.011	0.528	0.028	0.012
30	0.584	0.033	0.014	0.608	0.036	0.018	0.633	0.040	0.017
35	0.682	0.043	0.019	0.709	0.048	0.021	0.739	0.053	0.023
40	0.779	0.056	0.024	0.810	0.061	0.026	0.844	0.068	0.029
45	0.877	0.069	0.030	0.911	0.076	0.033	0.950	0.084	0.036
50	0.974	0.084	0.036	1.013	0.092	0.040	1.055	0.102	0.044
75	1.461	0.178	0.077	1.519	0.196	0.085	1.583	0.218	0.094
100	1.948	0.303	0.131	2.025	0.333	0.144	2.111	0.369	0.160
125	2.435	0.459	0.199	2.532	0.504	0.218	2.639	0.557	0.241
150	2.922	0.643	0.278	3.038	0.706	0.306	3.166	0.781	0.338
175	3.409	0.855	0.370	3.544	0.940	0.407	3.694	1.039	0.450
200	3.897	1.095	0.474	4.051	1.204	0.521	4.222	1.331	0.576
250	4.871	1.655	0.717	5.063	1.819	0.788	5.277	2.012	0.871
300	5.845	2.320	1.005	6.076	2.550	1.104	6.333	2.820	1.222
350	6.819	3.087	1.337	7.089	3.393	1.469	7.388	3.752	1.625
400	7.793	3.953	1.712	8.101	4.345	1.882	8.444	4.806	2.083
450	8.767	4.917	2.130	9.114	5.404	2.340	9.499	5.976	2.588
500	9.741	5.976	2.588	10.127	6.568	2.845	10.555	7.264	3.146
600	11.690	8.377	3.628	12.152	9.206	3.987	12.666	10.182	4.410
700	13.638	11.145	4.827	14.178	12.248	5.305	14.777	13.546	5.867

(continued on next page)

152 PVC PIPE—DESIGN AND INSTALLATION

Table B-4 Flow friction loss, AWWA C909 PVCO pipe (continued)

Flow (gpm)	6 in. CIOD (AWWA C909) CL 100			6 in. CIOD (AWWA C909) CL 150			6 in. CIOD (AWWA C909) CL 200		
	Velocity (ft/sec)	Loss of Head (ft/100 ft)	Pressure Drop (psi/100 ft)	Velocity (ft/sec)	Loss of Head (ft/100 ft)	Pressure Drop (psi/100 ft)	Velocity (ft/sec)	Loss of Head (ft/100 ft)	Pressure Drop (psi/100 ft)
50	0.471	0.014	0.006	0.490	0.016	0.007	0.511	0.018	0.008
60	0.565	0.020	0.009	0.588	0.022	0.010	0.613	0.025	0.011
70	0.659	0.027	0.012	0.686	0.029	0.013	0.716	0.033	0.014
75	0.706	0.030	0.013	0.736	0.034	0.015	0.767	0.037	0.016
80	0.753	0.034	0.015	0.785	0.038	0.016	0.818	0.042	0.018
90	0.848	0.043	0.018	0.883	0.047	0.020	0.920	0.052	0.023
100	0.942	0.052	0.022	0.981	0.057	0.025	1.022	0.063	0.027
125	1.177	0.078	0.034	1.226	0.086	0.037	1.278	0.096	0.041
150	1.413	0.110	0.047	1.471	0.121	0.052	1.534	0.134	0.058
175	1.648	0.146	0.063	1.716	0.161	0.070	1.789	0.178	0.077
200	1.884	0.187	0.081	1.961	0.206	0.089	2.045	0.228	0.099
250	2.354	0.282	0.122	2.452	0.312	0.135	2.556	0.345	0.149
300	2.825	0.396	0.171	2.942	0.437	0.189	3.067	0.483	0.209
350	3.296	0.527	0.228	3.432	0.581	0.252	3.578	0.643	0.279
400	3.767	0.674	0.292	3.923	0.744	0.322	4.090	0.824	0.357
450	4.238	0.839	0.363	4.413	0.926	0.401	4.601	1.024	0.444
500	4.709	1.020	0.442	4.903	1.125	0.487	5.112	1.245	0.539
600	5.651	1.429	0.619	5.884	1.577	0.683	6.134	1.745	0.756
700	6.592	1.901	0.823	6.865	2.098	0.909	7.157	2.322	1.006
800	7.534	2.435	1.054	7.846	2.687	1.164	8.179	2.973	1.288
1,000	9.418	3.681	1.594	9.807	4.062	1.759	10.224	4.495	1.947
1,500	14.127	7.799	3.378	14.710	8.606	3.727	15.336	9.524	4.125

(continued on next page)

APPENDIX B 153

Table B-4 Flow friction loss, AWWA C909 PVCO pipe (continued)

Flow (gpm)	8 in. CIOD (AWWA C909) CL 100			8 in. CIOD (AWWA C909) CL 150			8 in. CIOD (AWWA C909) CL 200		
	Velocity (ft/sec)	Loss of Head (ft/100 ft)	Pressure Drop (psi/100 ft)	Velocity (ft/sec)	Loss of Head (ft/100 ft)	Pressure Drop (psi/100 ft)	Velocity (ft/sec)	Loss of Head (ft/100 ft)	Pressure Drop (psi/100 ft)
100	0.547	0.014	0.006	0.570	0.015	0.007	0.594	0.017	0.007
125	0.684	0.021	0.009	0.713	0.023	0.010	0.743	0.026	0.011
150	0.821	0.029	0.013	0.855	0.032	0.014	0.891	0.036	0.015
175	0.958	0.039	0.017	0.998	0.043	0.019	1.040	0.048	0.021
200	1.095	0.050	0.022	1.140	0.055	0.024	1.188	0.061	0.026
250	1.369	0.075	0.033	1.425	0.083	0.036	1.485	0.092	0.040
300	1.642	0.106	0.046	1.710	0.117	0.051	1.782	0.129	0.056
350	1.918	0.141	0.061	1.995	0.155	0.067	2.079	0.172	0.074
400	2.190	0.180	0.078	2.280	0.199	0.086	2.376	0.220	0.095
450	2.464	0.224	0.097	2.565	0.247	0.107	2.674	0.273	0.118
500	2.737	0.272	0.118	2.850	0.301	0.130	2.971	0.332	0.144
600	3.285	0.382	0.165	3.420	0.421	0.183	3.565	0.466	0.202
700	3.832	0.508	0.220	3.991	0.561	0.243	4.159	0.620	0.268
800	4.380	0.651	0.282	4.561	0.718	0.311	4.753	0.794	0.344
1,000	5.475	0.984	0.426	5.701	1.085	0.470	5.941	1.200	0.520
1,200	6.570	1.379	0.597	6.841	1.521	0.659	7.129	1.682	0.728
1,400	7.665	1.834	0.794	7.981	2.024	0.877	8.318	2.238	0.969
1,600	8.760	2.349	1.017	9.121	2.592	1.122	9.506	2.866	1.241
2,000	10.949	3.551	1.538	11.402	3.918	1.697	11.882	4.332	1.876

(continued on next page)

154 PVC PIPE—DESIGN AND INSTALLATION

Table B-4 Flow friction loss, AWWA C909 PVCO pipe (continued)

Flow (gpm)	10 in. CIOD (AWWA C909) CL 100			10 in. CIOD (AWWA C909) CL 150			10 in. CIOD (AWWA C909) CL 200		
	Velocity (ft/sec)	Loss of Head (ft/100 ft)	Pressure Drop (psi/100 ft)	Velocity (ft/sec)	Loss of Head (ft/100 ft)	Pressure Drop (psi/100 ft)	Velocity (ft/sec)	Loss of Head (ft/100 ft)	Pressure Drop (psi/100 ft)
175	0.637	0.014	0.006	0.664	0.016	0.007	0.691	0.018	0.008
200	0.728	0.018	0.008	0.759	0.020	0.009	0.790	0.023	0.010
250	0.910	0.028	0.012	0.948	0.031	0.013	0.987	0.034	0.015
300	1.092	0.039	0.017	1.138	0.043	0.019	1.185	0.048	0.021
350	1.273	0.052	0.023	1.327	0.057	0.025	1.382	0.064	0.028
400	1.455	0.067	0.029	1.517	0.074	0.032	1.580	0.081	0.035
450	1.637	0.083	0.036	1.707	0.092	0.040	1.777	0.101	0.044
500	1.819	0.101	0.044	1.896	0.111	0.048	1.974	0.123	0.053
600	2.183	0.141	0.061	2.276	0.156	0.068	2.369	0.172	0.075
700	2.547	0.188	0.081	2.655	0.208	0.090	2.764	0.229	0.099
800	2.911	0.241	0.104	3.034	0.266	0.115	3.159	0.294	0.127
1,000	3.638	0.364	0.158	3.793	0.402	0.174	3.949	0.444	0.192
1,200	4.366	0.510	0.221	4.551	0.563	0.244	4.739	0.623	0.270
1,400	5.094	0.679	0.294	5.310	0.749	0.325	5.528	0.828	0.359
1,600	5.821	0.869	0.376	6.068	0.959	0.416	6.318	1.061	0.459
2,000	7.277	1.314	0.569	7.585	1.450	0.628	7.898	1.604	0.695
2,500	9.096	1.986	0.860	9.481	2.193	0.950	9.872	2.424	1.050
3,000	10.915	2.784	1.206	11.378	3.073	1.331	11.846	3.398	1.472

(continued on next page)

Table B-4 Flow friction loss, AWWA C909 PVCO pipe (continued)

	12 in. CIOD (AWWA C909) CL 100			12 in. CIOD (AWWA C909) CL 150			12 in. CIOD (AWWA C909) CL 200		
Flow (gpm)	Velocity (ft/sec)	Loss of Head (ft/100 ft)	Pressure Drop (psi/100 ft)	Velocity (ft/sec)	Loss of Head (ft/100 ft)	Pressure Drop (psi/100 ft)	Velocity (ft/sec)	Loss of Head (ft/100 ft)	Pressure Drop (psi/100 ft)
50	0.129	0.001	0.000	0.134	0.001	0.000	0.140	0.001	0.000
75	0.193	0.001	0.001	0.201	0.001	0.001	0.209	0.002	0.001
100	0.257	0.002	0.001	0.268	0.002	0.001	0.279	0.003	0.001
125	0.322	0.003	0.001	0.335	0.004	0.002	0.349	0.004	0.002
175	0.450	0.006	0.003	0.469	0.007	0.003	0.489	0.008	0.003
200	0.515	0.008	0.003	0.536	0.009	0.004	0.559	0.010	0.004
250	0.643	0.012	0.005	0.670	0.013	0.006	0.698	0.015	0.006
300	0.772	0.017	0.007	0.804	0.019	0.008	0.838	0.021	0.009
350	0.901	0.022	0.010	0.938	0.025	0.011	0.978	0.027	0.012
400	1.029	0.029	0.012	1.072	0.032	0.014	1.117	0.035	0.015
450	1.158	0.036	0.015	1.206	0.039	0.017	1.257	0.044	0.019
500	1.287	0.043	0.019	1.340	0.048	0.021	1.398	0.053	0.023
600	1.544	0.061	0.026	1.608	0.067	0.029	1.676	0.074	0.032
700	1.801	0.081	0.035	1.876	0.089	0.039	1.955	0.099	0.043
800	2.058	0.104	0.045	2.144	0.114	0.050	2.234	0.127	0.055
900	2.316	0.129	0.056	2.412	0.142	0.062	2.514	0.157	0.068
1,000	2.573	0.157	0.068	2.680	0.173	0.075	2.793	0.191	0.083
1,200	3.088	0.220	0.095	3.216	0.243	0.105	3.351	0.268	0.116
1,400	3.602	0.292	0.127	3.752	0.323	0.140	3.910	0.357	0.154
1,600	4.117	0.374	0.162	4.288	0.413	0.179	4.469	0.457	0.198
1,800	4.631	0.465	0.202	4.824	0.514	0.223	5.027	0.568	0.246
2,000	5.146	0.566	0.245	5.360	0.625	0.271	5.586	0.691	0.299

APPENDIX B   155

This page intentionally blank.

Bibliography

Abrasion Resistance, Das Kunststoffrohr V. 13 (25). (July 1969).

Airport Runway Depth of Cover Tables. National Corrugated Steel Pipe Association. Shiller Park, IL.

Arnold, G.E. Experience With Main Breaks in Four Large Cities. *Jour. AWWA* (Aug. 1960).

ANSI/AWWA C605 Standard for Underground Installation of PVC Pressure Pipe and Fittings for Water. American Water Works Association, Denver, CO (latest edition).

ANSI/AWWA C651 Standard for Disinfecting Water Mains. American Water Works Association, Denver, CO (latest edition).

ANSI/AWWA C800 Standard for Underground Service Line Valves and Fittings. American Water Works Association, Denver, CO (latest edition).

ANSI/AWWA C900 Standard for Polyvinyl Chloride (PVC) Pressure Pipe and Fabricated Fittings, 4 In. Through 12 In. (100 mm Through 300 mm), for Water Distribution. American Water Works Association, Denver, CO (latest edition).

ANSI/AWWA C905 Standard for Polyvinyl Chloride (PVC) Pressure Pipe and Fabricated Fittings, 14 In. Through 48 In. (350 mm Through 1200 mm), for Water Transmission and Distribution. American Water Works Association, Denver, CO (latest edition).

ANSI/AWWA C907 Standard for Polyvinyl Chloride (PVC) Pressure Fittings for Water-4 In. Through 8 In. (100 mm Through 200 mm). American Water Works Association, Denver, CO (latest edition).

ANSI/AWWA C908 Standard for PVC Self-Tapping Saddle Tees for Use on PVC Pipe. American Water Works Association, Denver, CO (latest edition).

ANSI/AWWA C909 Standard for Molecularly Oriented Polyvinyl Chloride (PVCO) Pressure Pipe, 4 In. Through 24 In. (100 mm Through 600 mm), For Water Distribution. American Water Works Association, Denver, CO (latest edition).

AWWA M17, *Installation, Field Testing, and Maintenance of Fire Hydrants.* American Water Works Association, Denver, CO (latest edition).

Babbitt, H.E. et al. *Water Supply Engineering.* (6th ed.) p. 574.

Barnard, R.E. Design and Deflection Control of Buried Steel Pipe Supporting Earth and Live Loads. American Society for Testing and Materials. Proc. 57 (1957).

Bishop, R.R. The Structural Performance of Polyvinyl Chloride Pipe Subjected to External Soil Pressures. Unpublished Master's Thesis, Utah State University. Logan, UT (May 1973).

Bishop, R.R., and R.W. Jeppson. Hydraulic Characteristics of PVC Pipe in Sanitary Sewers. Utah State University, Logan, UT (Sep. 1975).

Bulkey, C.W., R.G. Morin, and A.J. Stockwell. Vinyl Polymers and Copolymers. Modern Plastics Encyclopedia, 45:14A:336 (Oct. 1968).

Chang, F.S.C. Prediction of Long-Time Pipe Bursting Stress From Short-Time Tests. Society of Plastics Engineers 27th Annual Tech. Conf. Papers V15 (May 1969). p. 154.

Chemical Resistance Handbook. Plastiline Inc. Cat. PGF 0970-1, Pompano Beach, FL (1970).

Descriptions of Plastic Piping Joints. PPI Technical Note, PPI-TN 10. Plastics Pipe Institute, New York (Mar. 1975).

Design and Construction of Sanitary and Storm Sewers. ASCE Manual and Report on Engineering Practice No. 37 (WPCF Manual of Practice No. 9). American Society of Civil Engineers and the Water Pollution Control Federation. New York (1974).

Findley, W.N., and J.F. Tracy. 16-Year Creep of Polyethylene and PVC. MRLE-88, EMRL-57. Materials Sciences Program, Brown University, Providence, RI (Nov. 1973).

Flow of Fluids Through Valves, Fittings and Pipe. Technical Paper No. 410, 12th Printing. Crane Co., Chicago, IL (1972).

Handbook of Drainage and Construction Products. Armco Drainage and Metal Products, Inc., Middletown, OH (1955) p. 554.

Handbook of Steel Drainage and Highway Construction Products. American Iron and Steel Institute. Donnelley and Sons, Co. (1971).

Heilmayr, P.F. PVC Pipe Keeps Rolling Along. *Plastics Engineering* (Jan. 1976) p. 26.

Hendricks, J.C. Weathering Properties of Vinyl Plastics. *Plastics Technology* (Mar. 1955) p. 81.

Hendricks, J.C., and E.L. White. *Weathering Characteristics of Polyvinyl Chloride Type Plastics*. National Lead Company Research Laboratories, Brooklyn, NY, Wire and Wire Products (1952).

Henson, J.H.L., and A. Whelan. *Developments in PVC Technology*. National College of Rubber Technology, London (Feb. 1973).

Hermes, R.M. On the Inextensional Theory of Deformation of a Right Circular Cylindrical Shell. West Coast National Conf. of the Applied Mechanics Div., ASME (June 1951).

Hertzberg, L.G. Suggested Non-Technical Manual on Corrosion for Water Works Operators. *Jour. AWWA*, 48:6:719 (June 1956).

Howard, A.K. Laboratory Load Tests on Buried Flexible Pipe. *Jour. AWWA* (Oct. 1972).

Howard, A.K. Modulus of Soil Reaction (E') Values for Buried Flexible Pipe. *Jour. of the Geotechnical Engineering Division*, ASCE 103:GT. Proc. Paper 12700 (Jan. 1977).

Hucks, R.T. Designing PVC Pipe for Water-Distribution Systems. *Jour. AWWA*, 64:443 (1972).

Hucks, R.T. Design of PVC Water Distribution Pipe. *Civil Engineering*, ASCE, 42:6:70 (June 1972).

Jeppson, R.W. *Analysis of Flow in Pipe Networks*. Ann Arbor Science, Ann Arbor, MI (1977).

Jeppson, R.W., G.H. Flammer, and G.Z. Watters. Experimental Study of Water Hammer in Buried PVC and Permastran Pipes. Utah Water Research Laboratory/College of Engineering. Utah State University, Logan, UT (Apr. 1972).

Kennedy, Harold Jr., Dennis Shumard, and M. Cary. Investigation of Pipe-to-Soil Friction and Its Affect on Thrust Restraint Design for PVC and Ductile Iron Pipe. EBAA Iron Inc., *Presented at AWWA Distribution Systems Symposium*, September 1989.

Kennedy, Harold Jr., Dennis Shumard, and M. Cary. *PVC Pipe Thrust Restraint Design Handbook*. EBAA Iron Inc., Eastland, TX.

Kerr, S.L. Effect of Valve Action on Water Hammer. *Jour. AWWA*, 52:65 (1960).

Kern, Robert. How to Compute Pipe Size. *Chemical Engineering* (Jan. 1975). p. 115.

Kerr, S.L. Surges in Pipelines–Oil and Water. *Trans.* ASME, 72:667 (1950).

Kerr, S.L. Water Hammer–A Problem in Engineering Design. *Consulting Engineer* (May 1958).

Kerr, S.L. Water Hammer Control. *Journal AWWA*, 43:12:985 (Dec. 1951).

Killeen, N.D., and J.S. Schaul. Method of Determining Hydrostatic Design Stresses for PVC Pressure Pipe. *Interpace Technical Journal*, 1:1:17 (1964).

Kolp, D.A. Water Hammer Generated by Air Release. Colorado State University Thesis, Ft. Collins, CO (Aug. 1968).

Manual of Recommended Practice, American Railway Engineering Assoc., AREA Spec 1-4-28, Chicago, IL.

Manual on Sulfides in Sewers. US Environmental Protection Agency.

Marston, Anson, and A.O. Anderson. The Theory of Loads on Pipes in Ditches and Tests of Cement and Clay Drain Tile and Sewer Pipe. Bul. 31, Iowa, Iowa Engineering Experiment Station, Ames, IA (1913).

Modern Plastics Encyclopedia. Issued annually by Modern Plastics. McGraw-Hill, New York, NY.

Morrison, E.B. Nomograph for the Design of Thrust Blocks. Civil Engineering, ASCE (June 1969).

Moser, A.P., R.K. Watkins, and O.K. Shupe. Design and Performance of PVC Pipes Subjected to External Soil Pressure. Buried Structures Laboratory, Utah State University, Logan, UT (June 1976).

The Nature of Hydrostatic Time-to-Rupture Plots. PPI Technical Note PPI-TN 7. Plastics Pipe Institute, New York, NY (Sept. 1973).

Neale, L.C. and R.E. Price. Flow Characteristics of PVC Sewer Pipe. ASCE, *Jour. of Sanitary Engineers Div.* Proc. 90 SA3, 109 (1964).

Nesbeitt, W.D. Long-life Safety of PVC Water Pipe. Modern Plastics, New York, NY (Nov. 1975).

Nesbeitt, W.D. PVC Pipe in Water Distribution: Reliability and Durability. *Jour. AWWA,* 67:10:576 (Oct. 1975).

Newmark, N.M. Influence Charts for Computation of Stresses in Elastic Foundations. University of Illinois, Engineering Experiment Station, Bull. 338 (1942).

Parmakian, J. Pressure Surges at Large Pump Installations. *Trans.,* ASME, 75:995 (1953).

Parmakian, J. *Water Hammer Analysis.* Prentice-Hall Inc., New York, NY (1955).

Penn, W.S. *PVC Technology.* Wiley Interscience, a Division of John Wiley and Sons Inc., New York (1967).

Perry, J.H. *Chemical Engineer's Handbook.* McGraw-Hill, New York, NY (3rd ed.) p. 377.

Pipe Friction Manual. Hydraulic Institute, New York, NY (3rd ed., 1961).

Pipeline Design for Water and Wastewater. American Society of Civil Engineers, New York, NY (1975).

Plastics Piping Manual. Vol. 1. Plastics Pipe Institute, New York, NY (1976).

Policies and Procedures for Developing Recommended Hydrostatic Design Stresses for Thermoplastic Pipe Materials. PPI Technical Report, PPI-TR3. Plastics Pipe Institute, New York, NY (June 1975).

Poly (Vinyl Chloride) (PVC) Plastic Piping Design and Installation. PPI Technical Report PPI-TR13. Plastics Pipe Institute, New York, NY (Aug. 1973).

Potyondy, J.G. Skin Friction Between Various Soils and Construction Materials. *Geotechnique,* London, England, Volume II, No. 4 (Dec. 1961) pp. 339-353.

PVC Pipe—Technology Serving the Water Industry, Uni-Bell PVC Pipe Association, Dallas, TX (1984).

PVC Pipe for Water Distribution Systems. Technical Report, Information presented by the Plastics Pipe Institute to AWWA Standards Committee on Plastics Pipe, Chicago, IL (June 1972).

PVC, Plastics Engineering Primer. *Plastics Engineering,* 29:12:25 (Dec. 1973).

PVC Resins and Compounds, Allied Chemical Technical Bull. Allied Chemical Corporation (Oct. 1972).

Recommendations for Storage and Handling of Polyvinyl Chloride Plastic (PVC) Pipe. PPI Technical Report, PPI-TR26. Plastics Pipe Institute, New York, NY (May 1975).

Recommended Performance Specification for Joint Restraint Devices for Use With Polyvinyl Chloride (PVC) Pipe, UNI-B-13. Uni-Bell PVC Pipe Association, Dallas, TX (1992).

Recommended Practice for the Installation of Polyvinyl Chloride (PVC) Pressure Pipe, UNI-B-3. Uni-Bell PVC Pipe Association, Dallas, TX (1992).

Recommended Practice for Making Solvent Cemented Joints with Polyvinyl Chloride Plastics (PVC) Pipe and Fittings. PPI Technical Report, PPI-TR10. Plastics Pipe Institute, New York, NY (Feb. 1969).

Recommended Service (Design) Factors for Pressure Applications of Thermoplastic Pipe Materials. PPI Technical Report, PPI-TR9. Plastics Pipe Institute, New York, NY (Aug. 1973).

Recommended Specification for Thermoplastic Pipe Joints, Pressure and Non-pressure Applications, UNI-B-1. Uni-Bell PVC Pipe Association, Dallas, TX (1990).

Reedy, D.R. Corrosion in the Water Works Industry. Materials Protection (Sept. 1966). p. 55.

Reinhart, F.W. Long-Term Hydrostatic Strengths of Thermoplastic Pipe. Proc. 4th American Gas Association Plastic Pipe Symposium, Arlington, VA (1973).

Reinhart, F.W. Long-Term Working Stress of Thermoplastic Pipe. *SPE Jour.,* 17:8:75 (Aug. 1961).

Reissner, E. On Finite Bending of Pressurized Tubes, *Jour. of Applied Mechanics Transactions,* ASME (Sept. 1959). p. 386.

Relative Abrasion Resistance of Ring-Tite™ PVC Pipe. Johns-Manville Sales Corporation, Long Beach, CA (Apr. 1972).

Resistance of Thermoplastic Piping Materials to Micro- and Macro-Biological Attack. PPI Technical Report, PPI-TR11. Plastics Pipe Institute, New York, NY (Feb. 1969).

Sansone, L.F. A Comparison of Short-Time Versus Long-Time Properties of Plastic Pipe Under Hydrostatic Pressure. *SPE Jour.,* 15:5:418 (May 1959).

Sarvetnick, H.A. *Polyvinyl Chloride*. Van Nostrand Reinhold Co., New York, NY (1969).

Spangler, M.G. The Structural Design of Flexible Pipe Culverts. Bull. 153, Iowa Engineering Experiment Station, Ames, IA (1941).

Spangler, M.G., and R.L. Handy. *Soil Engineering*, Intext Educational Publ., New York, NY (1973).

Standard Classification of Soils for Engineering Purposes, ASTM D2487. American Society for Testing and Materials, Philadelphia, PA (1993).

Standard Method of Test for Density of Soil and Soil-aggregate in Place by Nuclear Methods (Shallow Depth), ASTM D2922. American Society for Testing and Materials, Philadelphia, PA (1976).

Standard Method of Test for Density of Soil in Place by the Rubber-Balloon Method, ASTM D2167. American Society for Testing and Materials, Philadelphia, PA (1977).

Standard Method of Test for Density of Soil in Place by the Sand-Cone Method, ASTM D1556. American Society for Testing and Materials, Philadelphia, PA (1970).

Standard Method of Test for Moisture-Density Relations of Soils Using 5.5 lb (2.5 kg.) Rammer and 12 in. (204.8 mm) Drop, ASTM D698. American Society for Testing and Materials, Philadelphia, PA (1977).

Standard Method of Test for Relative Density of Cohesionless Soils, ASTM D2049. American Society for Testing and Materials, Philadelphia, PA (1969).

Standard No. 61 Drinking Water System Components-Health Effects. NSF International, Ann Arbor, MI (1992).

Standard Practice for Description of Soils (Visual-Manual Procedure), ASTM D2488. American Society for Testing and Materials, Philadelphia, PA (1993).

Standard Practice for Making Solvent-Cemented Joints with Poly (Vinyl Chloride) (PVC) Pipe and Fittings, ASTM D2855. American Society for Testing and Materials, Philadelphia, PA (1993).

Standard Practice for Underground Installation of Thermoplastic Pressure Piping, ASTM D2774. American Society for Testing and Materials, Philadelphia, PA (1972).

Standard Specification for Elastomeric Seals (Gaskets) for Joining Plastic Pipe, ASTM F477. American Society for Testing and Materials, Philadelphia, PA (1993).

Standard Specification for Highway Bridges. American Association of State Highway Officials. Washington, DC (1969).

Standard Specification for Joints for IPS PVC Pipe Using Solvent Cement, ASTM D2672. American Society for Testing and Materials. Philadelphia, PA (1993).

Standard Specification for Joints for Plastic Pressure Pipes Using Flexible Elastic Seals, ASTM D3139. American Society for Testing and Materials, Philadelphia, PA (1989).

Standard Specification for Poly (Vinyl Chloride) (PVC) Plastic Pipe (SDR Series), ASTM D2241. American Society for Testing and Materials, Philadelphia, PA (1993).

Standard Specification for Poly (Vinyl Chloride) Resins, ASTM D1755. American Society for Testing and Materials, Philadelphia, PA (1977).

Standard Specification for Public Works Construction. American Public Works Association and Associated General Contractors, Los Angeles, CA.

Standard Specification for Rigid Poly (Vinyl Chloride) (PVC) Compounds and Chlorinated Poly (Vinyl Chloride) (CPVC) Compounds, ASTM D1784. American Society for Testing and Materials, Philadelphia, PA (1992).

Standard Test Method for Degree of Fusion of Extruded Poly (Vinyl Chloride) Pipe and Molded Fittings by Acetone Immersion, ASTM D2152. American Society for Testing and Materials, Philadelphia, PA (1986).

Standard Test Method for Determination of the Impact Resistance of Thermoplastic Pipe and Fittings by Means of a Tup (Falling Weight), ASTM D2444. American Society for Testing and Materials, Philadelphia, PA (1993).

Standard Test Method for Determining Dimensions of Thermoplastic Pipe and Fittings, ASTM D2122. American Society for Testing and Materials, Philadelphia, PA (1990).

Standard Test Method for External Loading Properties of Plastic Pipe by Parallel-Plate Loading, ASTM D2412. American Society for Testing and Materials, Philadelphia, PA (1993).

Standard Test Method for Obtaining Hydrostatic Design Basis for Thermoplastic Pipe Materials, ASTM D2837. American Society for Testing and Materials, Philadelphia, PA (1992).

Standard Test Method for Short-Time Hydraulic Failure Pressure of Plastic Pipe, Tubing, and Fittings, ASTM D1599. American Society for Testing and Materials, Philadelphia, PA (1988).

Standard Test Method for Time-to-Failure of Plastic Pipe Under Constant Internal Pressure, ASTM D1598. American Society for Testing and Materials, Philadelphia, PA (1986).

Streeter, V.L. *Fluid Mechanics,* McGraw-Hill, New York, NY (2nd ed. 1958). p. 175.

Streeter, V.L., and E.V. Wylie. *Hydraulic Transients.* McGraw-Hill, New York, NY (1967).

Sudrabin, L.P. Protect Pipes From External Corrosion. *The American City and County* (May 1956). p. 65.

Symons, G.E. *Design and Selection: Valves, Hydrants, and Fittings.* Manual of Practice No. 4. Water and Wastes Engineering. Dun-Donnelley Publishing Corporation. New York, NY (May 1968).

Symons, G.E. *Water Systems Pipes and Piping.* Manual of Practice No. 2. Water and Wastes Engineering. Dun-Donnelley Publishing Corporation, New York, NY (May 1976).

Test to Determine Effect of an Undersized (smaller diameter than inside diameter of pipe and fittings) Electrical Sewer Pipe Auger on Schedule 40 PVC-1 Drain, Waste, and Vent Pipe and Fittings. Report from Research Laboratory, Carlon. Aurora, OH (May 1963).

Thermal Expansion and Contraction of Plastic Pipe. PPI Technical Report, PPI-TR21. Plastics Pipe Institute, New York, NY (Sept. 1973).

Thermoplastic Piping for the Transport of Chemicals. PPI Technical Report, PPI-TR19. Plastics Pipe Institute, New York, NY (Aug. 1973).

Thermoplastic Water Piping Systems. PPI Technical Report, PPI-TR16. Plastics Pipe Institute, New York, NY (Aug. 1973).

Tiedeman, W.D. A Study of Plastic Pipe for Potable Water Supplies. National Sanitation Foundation, Ann Arbor, MI (1955).

Timoshenko, S.P. *Strength of Materials, Part 11-Advanced Theory and Problems,* Van Nostrand Co., Princeton, NJ (1968). p. 187.

Timoshenko, S.P. *Theory of Elastic Stability,* McGraw-Hill (2nd ed. 1961).

Timoshenko, S.P., and D.H. Young. *Elements of Strength of Materials.* Van Nostrand Co., Princeton, NJ (4th ed.). p. 111, p. 139.

Tipps, C.W. Underground Corrosion. *Materials Protection.* (Sept. 1966). p. 9.

Tobin, W.W. Stabilization of Rigid Polyvinyl Chloride Against Ultraviolet Radiation. Presented at Society of Plastic Engineers 21st Annual Technical Conf. Boston, MA (Mar. 1965).

Transport fester Stoffe durch PVC-hart—Rohre (Transport of Solid Substances Through Hard-PVC Pipes). Code: 237-4032-1, German.

Uni-Bell Handbook of PVC Pipe-Design and Construction. Uni-Bell PVC Pipe Association, Dallas, TX (3rd ed.) (1991).

Water Flow Characteristics of Thermoplastic Pipe. PPI Technical Report, PPI-TR14. Plastics Pipe Institute, New York, NY (Mar. 1971).

Watkins, R.K. Design of Buried, Pressurized Flexible Pipe. ASCE National Transportation Engineering Meeting in Boston, MA. Appendix C (July 1970).

Watkins, R.K., and A.P. Moser. Response of Corrugated Steel Pipe to External Soil Pressures. Highway Research Record 373 (1971). p. 88.

Watkins, R.K., A.P. Moser, and R.R. Bishop. Structural Response of Buried PVC Pipe. Modern Plastics (Nov. 1973). p. 88.

Watkins, R.K., and A.B. Smith. Ring Deflection of Buried Pipe. *Jour. AWWA,* 59:3 (Mar. 1967).

Watkins, R.K., and M.G. Spangler. Some Characteristics of the Modulus of Passive Resistance of Soil-A Study in Similitude.

Watters, G.Z. The Behavior of PVC Pipe Under the Action of Water Hammer Pressure Waves. Utah State University, Utah Water Research Laboratory Report, PRWG-93 (Mar. 1971).

Wear Data of Different Pipe Materials at Sewer Pipelines. The Institute for Hydromechanic and Hydraulic Structures, Technical University of Darmstadt, Darmstadt, W. Germany (May 7, 1973).

Wesfield, L.B., G.A. Thacker, and L.I. Nass. Photodegradation of Rigid Polyvinyl Chloride, *SPE Jour.*, 21:7:649 (July 1965).

White, H.C., and J.P. Layer. The Corrugated Metal Conduit As a Compression Ring. Highway Research Board Proceedings. Vol. 39 (1960). p. 389.

Wilging, R.C. Stress Rupture Testing of PVC Pipe, *Modern Plastics,* 57:10:90 (Oct. 1974).

Winding, C.C., and G.D. Hiatt. *Polymeric Materials*. McGraw-Hill, New York, NY (1961).

Wolter, F. Effect of Outdoor Weathering on the Performance of Some Selected Plastic Piping Materials. Presented by Battelle at the American Gas Association Fifth Plastic Pipe Symposium, Houston, TX (Nov. 1974).

Yearbook and Directory. The Los Angeles Rubber Group, Inc., Los Angeles, CA (1970).

Index

NOTE: An *f.* following a page number refers to a figure. A *t.* refers to a table.

Abrasion, 7-8
Allowable bending stress, 34-35, 38 *t.*
Arc test for fabricated fittings, 11
Assurance testing, 11-12
ASTM Specification D1784, 2
AWWA standards, 2

Bedding coefficient, 28, 28*f.*, 28*t.*
Bending
    ovalization, 38-39
    strain, 38, 38*t.*
    stress, 34-35, 38 *t.*
Biological attack, resistance to, 6-7
Boussinesq theory, 22

Casings, 86
    grouting pressures, 88, 88*t.*
    size recommendations, 86, 87*t.*
    skids, 86, 86*f.*
    spacers, 86, 87*f.*
Chemical evaluations, 2, 3*t.*
Chemical resistance
    gaskets, 4-5, 114 *t.*-127*t.*
    pipe, 4, 109*t.*-113*t.*
Control valves, 88, 90
Corrosion resistance, 2

Darcy-Weisbach equation, 13-14, 15 *f.*, 16*f.*
Deflection lag, 27-29
    factor, 27-28
Design stress, 57-58
Dimension measurement, 10-11
Dimension ratio, 59
Direct tapping, 99, 100*f.*
    corporation stops, 101
    coupon condition, 103-104, 104 *f.*
    cutting and tapping, 100-101, 101 *f.*, 102-104, 103 *f.*
    dry, 105
    mounting the machine, 102, 103*f.*
    nomenclature, 100*f.*
    outside pipe diameters, 102, 102*f.*
    procedure, 101-104
    tapping machines, 99-100
Disinfection, 96
    initial cleaning, 94

Distribution mains
    designing, 58-59
    disinfection, 94, 96
DR. *See* Dimension ratio

Earth loads. *See* Superimposed loads
Elastomeric seals, 7
Embedment
    materials and types, 84, 85*f.*
    procedure, 84-86
Environmental factors, effects of, 5-8
Extrusion compounds, 1-2
Extrusion quality test, 11

Fabricated-fitting pressure test, 11
Fire hydrants, 88, 90, 90*f.*
Fittings, 88
    assembly, 89
    backfill around, 89
    fabricated, 11, 63, 89
    handling, 89
    injection-molded, 62-63, 89
Flattening test, 11
Flexible pipe, 24-26
    defined, 24
    deflection lag, 27-29, 28 *f.*, 28*t.*, 29*t.*, 30*t.*, 31*t.*, 32*f.*, 32*t.*
    design example, 29-33
    pipe stiffness, 24-25, 26 *t.*
    Spangler's Iowa deflection formula, 26-27
Flow
    capacity, 17
    coefficients (C), 17
    formulas, 13-18
    rate, 14-17
    velocity, 14, 17
Friction factor, 14, 15*f.*
Friction loss, 17, 18*f.*, 19*f.*, 69
    ASTM D2241/AWWA C905 pipe, 144*t.*-150*t.*
    AWWA C900 PVC pipe, 130*t.*-134*t.*
    AWWA C905 pipe, 135*t.*-143*t.*
    AWWA C909 PVCO pipe, 151*t.*-155*t.*
Gasketed joint design testing, 9

Hazen–Williams equation, 14–17
HDB. *See* Hydrostatic design basis
Head loss, 17, 18*f.*, 19*f.*
Hydrostatic and leakage testing, 94–95.
   See also Pressure testing
    allowable leakage, 96, 97*t.*
    duration, 95
    initial testing, 93
    preparations, 94
    test pressure, 95
    timing of, 93–94
Hydrostatic design basis, 57–58
Hydrostatic pressure
    and creep, 54–55
    and operating temperature, 53, 54*t.*
    quick-burst hoop stresses, 57, 57*f.*
    and stress, 54–58, 55 *t.*, 56*f.*, 57*f.*
    time dependence, 53–54
Hydrostatic proof test, 11–12

Inspection (when receiving shipment),
   73–74. *See also* Testing and inspection (in
   manufacturing)
Installation, 77
    alignment and grade, 77
    appurtenances, 88–91
    casings, 86–88
    joint restraint, 90
    pipe cutting and bending, 83–84
    pipe embedment, 84–86
    pipe joint assembly and offset, 82–83
    in trenches, 77–82
Iowa deflection formula, 26–27

Joint restraint, 47, 90, 91, 92*f. See also*
   Thrust blocks, Thrust restraint
    gasketed joint design testing, 9
    offset, 34, 37–38, 37 *f.*
    performance testing, 10
    restrained, 47

Lap-shear test, 10
Leakage testing. *See* Hydrostatic and leak-
   age testing
Live loads, 23, 23*f.*, 24*t.*
Long-term hydrostatic strength testing, 10
Longitudinal bending, 33–39, 36 *f.*
    allowable bending stress, 34–35, 38 *t.*
    bending ovalization, 38–39
    bending strain, 38, 38*t.*
    and installation, 84
    joint offset, 34, 37–38, 37 *f.*
Lubricants, 7

Maintenance program, 96–98
Marking inspection, 10
Marston's load theory, 21–22
Modified Iowa formula. *See* Iowa deflection
   formula
Modulus of soil reaction, 29, 29*t.*, 30*t.*, 31*t.*,
   32*t.*
Molecularly oriented polyvinyl chloride.
   See PVCO
Moody diagrams, 14, 15*f.*, 16*f.*

Permeation, 4
Pipe stiffness, 24–25, 26 *t.*
Point source loads, 22
Polyvinyl chloride, 1
    material properties, 1–2, 3 *t.*, 4*f.*
PR. *See* Pressure rating
Pressure fittings. *See* Fittings
Pressure rating, 60, 60*t.*
Pressure testing
    fabricated fitting, 11
    sustained pressure test, 11
Prism load, 28
Product packaging inspection, 11
PVC extrusion compound cell classification
   testing, 9
PVC pipe
    appurtenances, 88–91
    history, 1
    manufacturing sequence, 9
    pressure capacity, 53
    pressure classes, 59*t.*
    pressure ratings, 60*t.*
    short-term ratings, 61*t.*
    short-term strengths, 61*t.*
    unloading, inspection, storage, 73–76, 75 *f.*
PVCO, 1

Qualification testing, 9–10
Quality control testing, 10–11
Quick-burst test, 11

Rankine Passive Pressure formula, 51
Receiving, 73–74
Relative roughness, 14, 16*f.*
Resing, 1–2

Saddle tapping, 105, 106*f.*, 107*f.*
    corporation stop, 106
    cutting tool, 106
    procedure, 106
    service clamps or saddles, 105
    tapping machine, 105

Safety valves, 88
Service connections, 99
    direct tapping, 99–105
    saddle tapping, 105–107
    tapping sleeves and valves, 107–108
Soil
    bearing capacity, 44, 45t.
    bearing resistance, 48, 50t., 51–52, 51 t.
    classification chart, 31t.
    friction, 49–51, 50 f., 50t.
    internal shear strength, 49
    modulus of soil reaction, 29, 29t., 30t., 31t., 32t.
    plasticity chart, 32f.
    support combining factor, 29, 29t.
Spangler's Iowa deflection formula. *See* Iowa deflection formula
SR. *See* Stress regression *entries*
Standards, 2, 3t., 4f.
Static earth loads, 21–22
Storage, 75–76
Stress regression curve, 55, 55t.
Stress regression line, 55–56, 56 f.
Superimposed loads, 21–24, 26
Surge pressure, 63–64
    control techniques, 66–67
    cyclic surges, 64, 66
    transient surges, 64–65, 66 t.
Sustained pressure test, 11

Tapping sleeves and valves, 107–108
    cutting tool, 108
    procedure, 108
    tapping machine, 108
Taste-and-odor evaluations, 2, 3t.
Testing and inspection (in manufacturing). *See also* Hydrostatic and leakage testing, Inspection (when receiving shipment)
    assurance, 11–12
    hydrostatic proof, 11–12
    qualification, 9–10
    quality control, 10–11
Thermal effects on strength, 5, 6f.
Thermal expansion and contraction, 5–6, 40–42
    coefficients, 40, 40t.
    and length variation, 40, 40t., 41f.

Thrust blocks, 43, 91, 91f. *See also* Joint restraint
    design criteria, 43–45
    design examples, 45–47
    forces at a horizontal bend, 43–44, 44 f.
    gravity thrust blocks, 45
    installation, 90
    soil bearing capacity, 44, 45t.
Thrust restraint, 42–43
    bearing resistance, 48, 50t., 51–52, 51 t.
    pipe to soil friction, 49–51, 50 f., 50t.
    resistance to unbalanced thrust force, 48–51, 48 f.
    restrained joints, 47, 90, 91, 92f.
    thrust blocks, 43–47, 44 f., 45t.
Toxicological testing, 9
Transmission mains
    design considerations, 59–62
    design example, 67–72, 67 f.
    disinfection, 94, 96
Trenches, 77–78, 78 f.
    bedding, 78
    dewatering, 80–81
    embankments, 79
    excavating, 80
    final backfill, 78, 88
    foundation, 78, 81–82
    haunching, 78
    initial backfill, 78
    laying of pipe, 82
    stockpiling excavated materials, 81
    subditches, 79, 79f.
    terminology, 78, 78f.
    transition width, 78–79
    width, 78–80, 79 f.
Tuberculation, 8

Ultraviolet degradation, 7
Uniformly distributed loads, 22
Unloading, 74, 75f.

Water hammer. *See* Surge pressure
Weathering resistance, 7
Workmanship inspection, 10

This page intentionally blank.

AWWA Manuals

M1, *Principles of Water Rates, Fees, and Charges*, Fifth Edition, 2000, #30001PA

M2, *Instrumentation and Control*, Third Edition, 2001, #30002PA

M3, *Safety Practices for Water Utilities*, Sixth Edition, 2002, #30003PA

M4, *Water Fluoridation Principles and Practices*, Fourth Edition, 1995, #30004PA

M5, *Water Utility Management Practices*, First Edition, 1980, #30005PA

M6, *Water Meters—Selection, Installation, Testing, and Maintenance*, Fourth Edition, 1999, #30006PA

M7, *Problem Organisms in Water: Identification and Treatment*, Second Edition, 1995, #30007PA

M9, *Concrete Pressure Pipe*, Second Edition, 1995, #30009PA

M11, *Steel Pipe—A Guide for Design and Installation*, Fourth Edition, 1989, #30011PA

M12, *Simplified Procedures for Water Examination*, Second Edition, 1997, #30012PA

M14, *Recommended Practice for Backflow Prevention and Cross-Connection Control*, Second Edition, 1990, #30014PA

M17, *Installation, Field Testing, and Maintenance of Fire Hydrants*, Third Edition, 1989, #30017PA

M19, *Emergency Planning for Water Utility Management*, Fourth Edition, 2001, #30019PA

M21, *Groundwater*, Second Edition, 1989, #30021PA

M22, *Sizing Water Service Lines and Meters*, First Edition, 1975, #30022PA

M23, *PVC Pipe—Design and Installation*, First Edition, 1980, #30023PA

M24, *Dual Water Systems*, Second Edition, 1994, #30024PA

M25, *Flexible-Membrane Covers and Linings for Potable-Water Reservoirs*, Third Edition, 2000, #30025PA

M27, *External Corrosion—Introduction to Chemistry and Control*, First Edition, 1987, #30027PA

M28, *Cleaning and Lining Water Mains*, Second Edition, 2001, #30028PA

M29, *Water Utility Capital Financing*, Second Edition, 1998, #30029PA

M30, *Precoat Filtration*, Second Edition, 1995, #30030PA

M31, *Distribution System Requirements for Fire Protection*, Third Edition, 1998, #30031PA

M32, *Distribution Network Analysis for Water Utilities*, First Edition, 1989, #30032PA

M33, *Flowmeters in Water Supply*, Second Edition, 1997, #30033PA

M36, *Water Audits and Leak Detection*, Second Edition, 1999, #30036PA

M37, *Operational Control of Coagulation and Filtration Processes*, Second Edition, 2000, #30037PA

M38, *Electrodialysis and Electrodialysis Reversal*, First Edition, 1995, #30038PA

M41, *Ductile-Iron Pipe and Fittings*, First Edition, 1996, #30041PA

M42, *Steel Water-Storage Tanks*, First Edition, 1998, #30042PA

M44, *Distribution Valves: Selection, Installation, Field Testing, and Maintenance*, First Edition, 1996, #30044PA

M45, *Fiberglass Pipe Design*, First Edition, 1996, #30045PA

M46, *Reverse Osmosis and Nanofiltration*, First Edition, 1999, #30046PA

M47, *Construction Contract Administration*, First Edition, 1996, #30047PA

M48, *Waterborne Pathogens*, First Edition, 1999, #30048PA

M49, *Butterfly Valves: Torque, Head Loss, and Cavitation Analysis*, First Edition, 2001, #30049PA

M50, *Water Resources Planning*, First Edition, 2001, #30050PA

M51, *Air-Release, Air/Vacuum, and Combination Air Valves*, First Edition, 2001, #30051PA

To order any of these manuals or other AWWA publications, call the Bookstore toll-free at 1-(800)-926-7337.